Introduction
to
Liquid State Physics

Introduction
to
Liquid State Physics

Clive A. Croxton

Fellow, Jesus College,
Cambridge University

JOHN WILEY & SONS
London · New York · Sydney · Toronto

Library of Congress Cataloging in Publication Data:

Croxton, Clive A.

Introduction to liquid state physics.

1. Liquids.   2. Statistical mechanics.   I. Title.
QC145.2.C76   530.4'2   74-13153

ISBN 0 471 18933 2 (Cloth)
ISBN 0 471 18934 0 (Paper)

Printed in Great Britain by J. W. Arrowsmith Ltd.,
Winterstoke Road, Bristol, England

To
Don Lewis

Preface

Tamed, but not entirely domesticated, analytically intractible yet not to be coerced by approximation, liquid state physics has for a long time been the *enfant terrible* of the phase diagram. The interested non-specialist understandably feels something of a *persona non grata*, yet the range of application of liquid state physics is immense. This book was written in the belief that the theory of liquids is now ready for dissemination to a wider and more general audience—to chemists, metallurgists, biophysicists and engineers, not to mention the non-specialist physicist who also needs to know something of the liquid state. This is not to say, of course, that the subject does not still have its problems: irreversibility, phase transitions and transport phenomena in classical systems are still poorly understood, as are critical phenomena. Nevertheless, we attempt here to present short accounts on all these topics as we do on liquid metals and the technologically important liquid-crystal systems. Inevitably some selection has to be made, and it was felt that the more complex molecular systems such as the glasses, water and liquid mixtures could not be included on this occasion. The equilibrium theory of simple fluids is dealt with at length, together with the statistical thermodynamics of the free liquid surface, and extensive illustrative use is made throughout of the computer simulation schemes which are now central aids in the theoretical development of the subject.

This book is in no way intended as a research monograph: no attempt is made to provide extensive references to the literature, although each chapter is concluded with a brief general bibliography. This book represents a wider-ranging but elementary account of the theory presented in *LSP* (*Liquid State Physics—A Statistical Mechanical Introduction*, Cambridge University Press) to which the reader requiring more detailed information is generally referred.

It is a great pleasure to acknowledge with thanks the discussions and interest of David Tabor and Tom Faber, and to thank once again all those to whom I was indebted in the production of *LSP*, on the basis of which this book was written.

CLIVE A. CROXTON

Cavendish Laboratory
June 1974

Acknowledgements

I am indebted to the following for permission to reproduce figures appearing in this book:

Fig. 1.5(b) (private collection of J. D. Bernal); Figs. 2.5, 2.17, 2.18 (F. H. Ree and W. G. Hoover, *J. Chem. Phys.*, **40**, 939 (1964), *J. Chem. Phys.*, **49**, 3609 (1968)); Figs. 4.9, 10.4, 10.6, 10.15(a, b) (A. Rahman, *Phys. Rev.*, **136**, A405 (1964), A. Paskin and A. Rahman, *Phys. Rev. Lett.*, **16**, 300 (1966)); Figs. 9.10, 9.11, 9.12, 10.9, 10.10 (W. McMillan, *Phys. Rev.*, **A4**, 1238 (1971), *Phys. Rev.*, **138**, A422 (1965)); Fig. 8.7(b) (D. Gustafson, A. Mackintosh, D. Zaffarano, *Phys. Rev.*, **130**, 1455 (1963)); Fig. 7.14 (J. A. Barker, *J. Chem. Phys.*, **60**, 1976 (1964)); Fig. 7.9 (T. Hill, *J. Chem. Phys.*, **20**, 141 (1952)); Fig. 6.10 (B. Chu and W. Kao, *J. Chem. Phys.*, **42**, 2608 (1965)); Fig. 6.3 (E. A. Guggenheim, *J. Chem. Phys.*, **13**, 253 (1945)); Figs. 3.13, 3.14, 4.7 (S. A. Rice and J. Lekner, *J. Chem. Phys.*, **42**, 3559 (1965)); Fig. 3.19 (N. Ashcroft and J. Lekner, *Phys. Rev.*, **145**, 83 (1966)); Figs. 3.18, 4.1(b), 4.6, 4.8(b), 4.10(a, b), 4.11, 5.14(a, b), 5.15, 5.18 (M. Klein, *J. Chem. Phys.*, **39**, 1388 (1963), M. Klein and M. S. Green, *J. Chem. Phys.*, **39**, 1367 (1963)); Figs. 4.3, 4.4 (S. Kimm, D. Henderson and L. Oden, *J. Chem. Phys.*, **45**, 4030 (1966)); Fig. 4.2 (D. Henderson and L. Oden, *Mol. Phys.*, **10**, 405 (1966)); Fig. 2.8 (J. S. Rowlinson, p. 65), Figs. 5.9(a, b), 5.11 (W. W. Wood, p. 175, p. 222) and Fig. 6.4 (P. A. Egelstaff and J. W. Ring, p. 264) in *Physics of Simple Liquids* (Eds. H. N. V. Temperley, J. S. Rowlinson and G. S. Rushbrooke, North-Holland Publishing Co., Amsterdam, 1968); Fig. 4.13 (N. H. March, *Liquid Metals*, Pergamon, Oxford, 1968); Figs. 9.1, 9.2, 9.3, 9.7, 9.9 (D. Luckhurst, *Physics Bulletin*, p. 279, May (1972), Institute of Physics and the Physical Society); Fig. 3.7(a) O. B. Verbeke and W. Brems; Figs. 7.17(a), 7.19 D. W. G. White; Fig. 8.3 (T. E. Faber, *Theory of Liquid Metals*, Cambridge University Press, Cambridge, 1972); Fig. 8.5 (T. E. Faber in *The Physics of Metals* (Ed. J. M. Ziman), Cambridge University Press, Cambridge, 1969); Figs. 1.3, 2.7, 2.11, 2.15, 2.16 (J. O. Hirschfelder, C. F. Curtiss and R. B. Bird, *Molecular Theory of Gases and Liquids*, Wiley, New York, 1954); Fig. 11.6 (S. A. Rice and P. Gray, *The Statistical Mechanics of Simple Liquids*, Wiley Interscience, New York, 1965); Figs. 3.11, 3.30(a, b, c), 4.1(a), 7.12, 7.16, 7.18, 10.11, 10.13, 10.14 (C. A. Croxton, *Liquid State Physics—A Statistical Mechanical Introduction*, Cambridge University Press, Cambridge, 1974); Figs. 9.14, 9.15 (S. Marčelja, *Nature*, **242**, 143 (1973)); Fig. 6.6 (M. E. Fisher, *J. Math. Phys.*, **5**, 944 (1964)).

I am also indebted for permission to reproduce the following Tables:

Table 2.3 (J. O. Hirschfelder, C. F. Curtiss and R. B. Bird, *Molecular Theory of Gases and Liquids*, Wiley, New York, 1954); Tables 4.3, 7.2, 11.2 (C. A. Croxton, *Liquid State Physics—A Statistical Mechanical Introduction*, Cambridge University Press, Cambridge, 1974); Table 6.2 (P. A. Egelstaff and J. W. Ring in *Physics of Simple Liquids*, Eds. H. N. V. Temperley, J. S. Rowlinson and G. S. Rushbrooke, North-Holland Publishing Co., Amsterdam, 1968); Tables 7.1, 8.2 (T. E. Faber, *Theory of Liquid Metals*, Cambridge University Press, Cambridge, 1972).

Contents

CHAPTER 1

The Liquid State

Introduction

The statistical description of the structure and dynamics of liquids, in comparison with the solid and gas phases, is relatively incomplete. The fundamental difficulty is that the liquid state lacks any idealized abstraction such as the ideal gas or perfect solid, which can later form a basis for theoretical refinement. Unlike the solid or the gas where either configurational or kinetic processes dominate the description, in liquid state physics we are confronted with the full and general statistical problem in which there are both dynamical and configurational contributions to the total energy. Approximation, if not judicious, will lead to a description of either high-density gases or disordered high-temperature solids. Indeed, at one time considerable effort was devoted to the representation of liquids in these terms, but it is now known that fluids do not have a simple interpolated status between gas and solid, although features of both the adjacent phases can be detected. It appears that a direct *a priori* statistical mechanical attack is necessary, and this forms the foundation of the modern theory of liquids to be developed in this book.

Although we now have a rather clear understanding of most of the fundamental processes operating in dense fluids—order of magnitude calculations and better are now quite feasible for many quantities—this is not to say that the development of the field is complete. Some major outstanding problems remain relating to phase transitions and critical phenomena, irreversibility and transport processes, and in this book we devote chapters specifically to these topics.

The ultimate objective of a statistical description of matter is, of course, to relate the macroscopic observables to the details of the intermolecular potential operating between particles in the assembly. Whilst we have no 'ideal liquid' to form a basis model, assemblies of *idealized particles* interacting through some simple model interaction have an important theoretical status, even if the results are not immediately comparable to real physical systems. The simple forms of interaction generally considered are hard sphere, square well, Gaussian, etc., and whilst no more than caricatures of realistic interactions they do allow a welcome simplification in the analysis. The advent of fast electronic computers largely offsets the absence of any real physical data with which to compare the theoretical results: it is now perfectly feasible to *simulate* systems of ~ 1000 particles interacting through these idealized interactions. Indeed, the computer schemes have now adopted such a central role that they represent a buffer between theory and experiment, and current theoretical developments are about equally preoccupied with both simulated and real data. In many respects

1

the simulations are preferable in that the wealth of microscopic data relating to molecular trajectories and distributions is far greater and more direct than that obtainable by conventional experiment.

This is not to say that we are able to abandon comparison with experiment entirely: ultimately, of course, we must face real problems. It is simply that comparison with experiment is temporarily abandoned in the interests of theoretical development and expediency: we shall discuss these computer schemes in Chapter 10.

Solids, Liquids and Gases

The ability of liquids and solids to form a free bounding surface obviously distinguishes them from gases which will, of course, diffusively fill the accessible volume. Indeed, the coefficients of self-diffusion of liquids ($\sim 10^{-5} \, \text{cm}^2 \, \text{s}^{-1}$) and solids ($\sim 10^{-9} \, \text{cm}^2 \, \text{s}^{-1}$) are orders of magnitude below that of the gas. On the other hand, the viscosities of gases and liquids are some thirteen orders of magnitude lower than those of solids, and this we may easily understand in terms of the molecular processes of momentum exchange. Flow in a solid consists primarily in rupturing of bonds and propagation of dislocations and imperfections. The resistance of a solid to a sustained shear stress is therefore understood in configurational terms and, as usual in a solid, potential contributions dominate. In a liquid there is both molecular transport and configurational readjustment and the flow process is characterized by both configurational and kinetic processes, whilst in a gas, of course, the flow is understood purely in terms of kinetic transport. It is in this limited sense that liquids have an interpolated status between gas and solid. Gas and solid processes compounded in appropriate amounts might well yield a result in agreement with observed liquid values, but the physical processes would not of course be apparent from such purely empirical considerations.

Many of the characteristic features of solids, liquids and gases can be understood qualitatively by plotting the Maxwellian velocity distribution against the molecular interaction, as shown in Figure 1.1. All three have a component of the distribution in the unbound region of the interaction, corresponding to the development of a vapour phase. In the gas, almost the entire distribution is in the unbound state. The majority of particles in the solid distribution are energetically confined to the symmetrical region near the bottom of the well, and we understand at once the development of harmonic motions in the lattice, and at the same time the increase in anharmonic contributions with increasing temperature.

Again, the increased vapour pressure, and the strongly anharmonic motions in the bound liquid phase are apparent from Figure 1.1. As the critical point is approached so the system passes through the bound–unbound condition with correspondingly large fluctuations between tenuously bound aggregates of several thousand molecules continuously forming and disrupting. It is the scattering of light from these clusters which is responsible for the shimmering milky appearance of a critical fluid, *critical opalescence*.

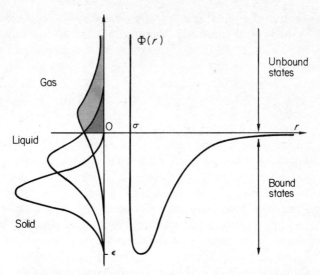

Figure 1.1. The Maxwell–Boltzmann energy distributions for a solid, liquid and gas plotted against a typical atomic pair potential. The shaded region represents the unbound component of the distribution. ε and σ are the well depth and atomic collision diameter, respectively

Whilst this qualitative description enables us to characterize some of the features of the solid, liquid and gaseous phases, what it does not tell us is why three states of matter exist, and why indeed we don't have one single phase whose features are simply characterized by the temperature. The theory of phase transitions is only poorly understood, and this very problem has been the focus of theoretical attack for some time now. The essential feature we have neglected so far, of course, is the *geometric* one concerning the packing of particles—the above description appealed only to the dynamic features of the assembly—known to be independent of the structure. Geometric considerations are largely irrelevant at gas densities, though the finite size of the molecular core does modify the ideal gas law as, for example, in van der Waals' equation of state. At liquid and solid densities geometric packing of spheres and the entropy differences between ordered and disordered arrangements intervenes to effect a lowering of the free energy through a modification of molecular order. This is of course a purely thermodynamic argument, and takes no account of the molecular processes involved in the phase transition. We shall see that the phase transition is essentially an *excluded volume* effect, the details of which will be presented in Chapter 5.

Classification of Liquids

Throughout this book we shall confine our attention to the so-called simple liquids—those characterized by spherically symmetric non-saturating interactions. Moreover, we shall assume that the forces are *central* in that they act through the centres of gravity, and that they are pairwise decomposable—that is, the total N-body configurational energy can be represented as the sum of pair interactions, where the factor $\frac{1}{2}$ prevents us counting the interactions ij and ji as distinct:

$$\Phi(1 \ldots N) = \frac{1}{2} \sum_{i}^{N} \sum_{j}^{N} \Phi(ij) = \sum_{i > j}^{N} \Phi(ij) \tag{1.1}$$

This is known not to be strictly true: three-body effects definitely intervene so that the total potential energy of three particles ijk is not simply $\Phi(ij) + \Phi(jk) + \Phi(ki)$, but is subject to a small *three-body* correction, $\Phi(ijk)$. The inclusion of these triplet potentials enormously complicates our task, and throughout this book we shall neglect three-body contributions, working in an essentially pair approximation. The resulting inaccuracy is likely to be negligible in comparison with the inherent approximation in the theoretical foundation of the subject itself.

Typical of this class of liquids are the liquid inert gases, and the various idealized interactions. A further subdivision as to whether these liquids exhibit quantum-mechanical features may be made: neon to a certain extent does, and helium can only be treated as a quantal fluid. If the de Broglie wavelength of the atom is of the order of the interparticle spacing we clearly have to abandon the classical, deterministic standpoint, and rigorously incorporate wave-mechanical effects.

This first classification may be extended to *symmetrical molecular systems* such as CH_4 (methane) which, whilst having an internal structure, has an effectively symmetrical central interaction by virtue of rotational 'sphericaliza-tion' so that neither molecule exerts a torque on the other. At high densities and low temperatures deviations from purely simple behaviour may be detected, but otherwise this appears a reasonable assumption.

The *liquid metals* are generally categorized as 'simple' systems in as far as their interactions are central, symmetrical and non-saturating with the possible exception of systems like germanium which show pronounced covalent ten-dencies towards bonding. Unlike the inert gases (except for three-body effects), the liquid-metal pseudopotential is temperature and density dependent, which leads to considerable complication in the calculation of thermodynamic and configurational properties. These systems are nevertheless included in this book, and discussed at some length in Chapter 8.

The *homonuclear diatomics* (N_2, O_2, H_2) and *heteronuclear diatomics of negligible dipole moment* (CO) cannot, of course, interact centrally or sym-metrically, but provided the internuclear separation is much smaller than the intermolecular spacing in the fluid the deviation from simple behaviour is not great. At low-temperature liquid densities there is a systematic departure from centro-symmetric systems, and various 'shape' corrections have been proposed to account for the deviation in terms of the configurational parameters of the molecule. We shall not consider these systems to any extent here.

The *lower hydrocarbons* show a pronounced discrepancy with respect to simple systems as the chain length becomes significant with respect to the intermolecular separation. We shall not consider these systems in this book, but the long rod-like molecules do form members of the class of *liquid crystals* which we shall mention again below.

The pronounced electrostatic interaction between simple but polar molecules such as SO_2 and molecules of great electrical and configurational asymmetry, such as water, are prohibitively difficult to handle by conventional statistical mechanical techniques in all but the most formal sense. We shall not discuss them in this book.

The *liquid-crystal* systems do not strictly lie within the scope of the present book, but their outstanding technological importance and physical interest is such that they have been arbitrarily included, and an elementary discussion is given in Chapter 9. These systems are characterized by the molecular requirement of a significant deviation from spherical symmetry, although only compounds with rod-like molecules seem to form liquid crystals. Further discussion of liquid-crystal systems is deferred to Chapter 9.

One of the more complex of the liquid classifications are the *glasses*—glycerol, rubber, glucose and other sugars in addition to ordinary glass. These systems are characterized by abnormally high viscosities in the vicinity of the freezing point which effectively 'hinder' the molecular rearrangement necessary for an ordered crystalline phase. In consequence the 'solid' phase lacks a crystalline structure and shows a typical conchoidal fracture pattern instead of the characteristic cleavage planes of a crystalline solid. One of the major distinctions generally drawn between solids and liquids is inability of the latter to sustain a shear stress. In fact this is not strictly true: if the molecular configuration cannot adjust to the stress rapidly enough, under the influence of a high-frequency stress for example, the mechanical response of the system will be more characteristic of a solid than a liquid, and this is the situation in glasses where the characteristic relaxation time may be of the order of hours or days. For very-low-frequency stresses, applied over periods of hours or years, the glasses exhibit non-Newtonian flow properties more characteristic of a liquid. Another example of this phenomenon is the 'bouncing silicone putty' which kneads like putty in the hand, yet bounces like rubber on the floor—again a question of rate of deformation. To return to the glasses, the specific heat of glycerol is shown in Figure 1.2. As the liquid cools, molecular translation and orientation is impeded to such an extent that the configurational adjustment necessary to crystallize does not occur and the system supercools along a metastable branch of the liquid. There are both translational and vibrational contributions to the specific heat along this branch, but at the *glass transformation temperature* T_g a 'freezing-out' or *ankylosis* of the translational degrees of freedom occurs, and the contributions are purely vibrational—similar in nature and magnitude to the crystal. The glass curve thus merges smoothly into the crystal curve at T_g, the only distinction between the amorphous glass phase and the crystalline solid being the extent of order.

6

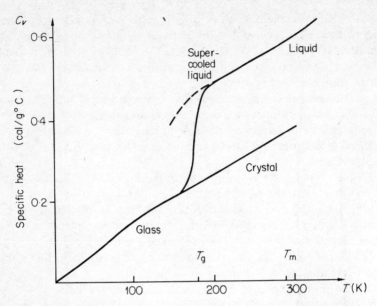

Figure 1.2 The specific heat of glycerol showing the crystalline and supercooled 'glassy' branches. The supercooled liquid branch merges into that of the amorphous solid at the glass transition temperature T_g, which is generally significantly less than the conventional melting temperature T_m

It is possible under certain circumstances to precipitate the liquid onto the crystalline branch—by raising the temperature well above T_g so that the molecular adjustment may proceed, for example—and this is devitrification. Devitrified glass has poor mechanical properties and readily shatters into small crystalline fragments.

The Potential Functions

Restricting the discussion to simple centro-symmetric non-saturating interactions from the outset, we need only consider one realistic interaction—the Lennard–Jones (6–12) potential—and a number of the idealized interactions, hard sphere, square well, etc. These potentials will prove adequate for our discussions in this book.

That a realistic interaction has both attractive and repulsive components is evident from the fact on one hand that solids and liquids obviously have the property of *cohesion*, but at the same time do not collapse indefinitely to a point singularity under the action of these forces. Evidently there must be a strongly repulsive core. The features of the interaction are established at an electronic level, and may be expressed formally as

$$\Phi(r) = u^{(val)} + u^{(ex)} + ar^{-6} + br^{-8} + cr^{-10} + \ldots \tag{1.2}$$

where $u^{(val)}$ is the valence energy of repulsion and $u^{(ex)}$ represents the repulsive exchange terms arising from Fermi prohibition of overlapping closed core states at small separations. The following terms are the dispersion energies arising from the dipole–dipole, dipole–quadrupole, quadrupole–quadrupole, . . . interactions respectively. The first explicit calculation for the interaction between two helium atoms was carried out by Slater and Kirkwood in 1931, and they obtained

$$\Phi(r) = \left\{ 770\, e^{-r/0\cdot217} - \frac{1\cdot49}{r^6} \right\} 10^{-12}\, erg \tag{1.3}$$

With regard to the theory of liquids it is desirable to have simpler, but still representative, expressions for the intermolecular potential since it generally appears in complicated integrals, and this can be achieved if a sufficiently high inverse power of separation is used for the repulsion. Thus we arrive at the Lennard–Jones model interaction

$$\Phi(r) = 4\varepsilon \left\{ \left(\frac{\sigma}{r}\right)^m - \left(\frac{\sigma}{r}\right)^n \right\} \tag{1.4}$$

where ε is the well depth and σ the collision diameter. The parameters m, n are generally taken as 12 and 6, although other combinations are frequently used. The other parameters, σ and ε, appearing in the Lennard–Jones 6–12 potential are generally determined experimentally to ensure an empirical fit to the data. A comparison of the Slater–Kirkwood He–He theoretical interaction and the LJ(6–12) function is made in Figure 1.3. It is seen that the LJ

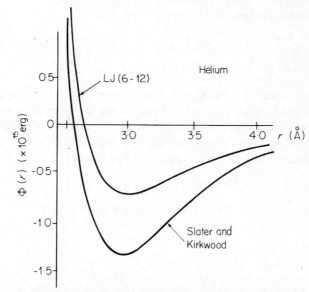

Figure 1.3 Comparison of the theoretical He–He interaction with the Lennard–Jones (6–12) model function parametrically fitted to virial data

8

potential is much flatter than the theoretical curve and leads to a larger equilibrium separation, r_0. The discrepancy is not as alarming as it might appear. The theoretical function is purely two-body, whilst the LJ function fitted to liquid data incorporates many-body effects which are known to modify the isolated two-body interaction. To this extent the LJ function is perhaps more appropriate to liquid systems, although it is of course weakly density dependent. Some values of the LJ(6–12) parameters are listed in Table 1.1.

Table 1.1. Lennard–Jones parameters for the inert gases

Gas	$\varepsilon \times 10^{-14}$ erg	σ (Å)
Helium	0·141	2·56
Neon	0·490	2·76
Argon	1·653	3·41
Krypton	2·290	3·66
Xenon	3·030	3·95

The idealized model interactions have a central role in the theoretical development and molecular dynamic simulation of dense liquids, and enable even rather complicated mathematical problems to be solved exactly or at least approximately. For this reason the *Gaussian* repulsive interaction is sometimes used, but its use is justified almost entirely on mathematical grounds, Gaussian integrals being rather easier to evaluate than most others.

The hard-sphere interaction

$$\left.\begin{aligned}\Phi(r) &= +\infty \qquad r \leqslant \sigma \\ &= 0 \qquad\quad r > \sigma \end{aligned}\right\} \tag{1.5}$$

embodies the effects of a strongly repulsive core, and since at liquid densities it is the core which establishes the main geometrical properties of the structure, we would expect this simplification to aid our theoretical understanding of the configurational processes operating in simple liquids.

Rather more realistic are the square-well systems:

$$\begin{aligned}\Phi(r) &= +\infty \qquad r \leqslant \sigma \\ &= -\varepsilon \qquad\ \sigma < r \leqslant g\sigma \\ &= 0 \qquad\quad r > g\sigma \end{aligned} \tag{1.6}$$

Now *attractive* contributions are involved, and indeed, the equation of state develops sinuous isotherms more characteristic of liquid systems. The *range* of the interaction g is generally taken as one or two atomic diameters: this provides us with an extra parameter to fit to the experimental data. In Table 1.2 we list values of σ, g and ε/k obtained from equation-of-state data for gases.

Table 1.2. Square well parameters

Gas	σ (Å)	g	ε/k K
Neon	2·38	1·87	19·5
Argon	3·16	1·85	69·4
Krypton	3·36	1·85	98·3
N_2	3·30	1·87	53·7
CO_2	3·92	1·83	119·0
CH_4	2·90	1·27	692
H_2O	2·61	1·20	1290

Structural Determinations

The 'intermediate' status of liquids between gases and solids is again observed in the X-ray and fast-neutron scattering experiments which provide reasonably direct structural information concerning the molecular organization in liquids. The liquid diffraction patterns show one or more diffuse rings or halos, whilst the solid phase shows either the bright Laue spot pattern characteristic of an extended ordered array, or alternatively sharp Debye–Scherrer rings in the case of polycrystalline systems (Figure 1.5). Gases, on the other hand, show a broad, structureless halo, and it was largely in terms of the polycrystalline solid and gas diffraction patterns that early attempts were made to 'explain' liquids as either dense, relatively ordered gases, or as disordered cybotactic solids. A number of variants on these models were proposed—virial extension of the equations of state to liquid densities, and cell models—but none of them proved adequate, and an *a priori* approach was started some years later which culminated in the modern statistical mechanical theory of liquids. Interestingly enough, it was Zernike and Prins' relation of the scattered X-ray intensity distribution to the microstructure which introduced the central concept of the equilibrium theory of liquids—the radial distribution function (RDF).

The intensity distribution at some distance away from the scattering assembly contains two distinct terms. There is the *independent* scattering from the N particles providing a background intensity, I_0. This is modulated by interference effects between the scattered wavefronts and this, of course, will depend on the configuration of the particles. It is the *modulation* of the intensity which is of importance and its measurement and inversion to yield the micro-structure represents the central objective of the elastic scattering experiments. The total scattered intensity may be easily shown to be

$$I(\theta) = I_0\left\{1 + \sum_{\substack{i,j=1\\i\neq j}}^{N} \frac{\sin sr_{ij}}{sr_{ij}}\right\}; \qquad s = \frac{4\pi \sin \theta}{\lambda} \qquad (1.7)$$

where the second term represents the fluctuation about the background intensity I_0, and this latter term is fully described by the parameters of the experimental set-up, the nature of the scattering atom, etc. The second term describes the interference of a first wave with a second, given that the two scattering centres are separated by the distance r_{ij}. Clearly this presupposes the molecular configuration. If the (continuous) radial distribution of scattering centres is $\rho_{(2)}(r)$ where ρ_L is the bulk liquid density,

$$\rho_{(2)}(r) = \rho_L g_{(2)}(r) \tag{1.8}$$

then the sum in the Debye formula (1.7) may be replaced by an integral

$$\sum_{\substack{i,j=1 \\ i \neq j}}^{N} \frac{\sin sr_{ij}}{sr_{ij}} \equiv 4\pi\rho_L \int_0^R g_{(2)}(r) \frac{\sin sr}{sr} r^2 \, dr \tag{1.9}$$

where R is an upper limit assumed larger than the size of the specimen. The final intensity distribution is then

$$I(\theta) = I_0 \left\{ 1 + 4\pi\rho_L \int_0^R g_{(2)}(r) \frac{\sin sr}{sr} r^2 \, dr \right\} \tag{1.10}$$

so that as $R \to \infty$, the fluctuation in the scattering intensity is related to the Fourier transform of the structure, $g_{(2)}(r)$. Of course when we come to *Fourier invert* the scattering data to obtain the distribution $g_{(2)}(r)$ there will be a statistical scatter in the data at large scattering angles, and this will introduce spurious ripples in the RDF. Again, scattering can only be detected out to some maximum scattering angle s_{max}, and *truncation errors* also introduce ripple into the inversion. It is for these and other reasons that the greatest care has to be exercised in the inversion process, and there is a marked preference to work theoretically in k-space, eliminating any spurious features inadvertently introduced in the transformation procedure.

Nevertheless, the distribution function shows all the qualitative features we might expect in an isotropic simple fluid (Figure 1.4). The radius vector whose

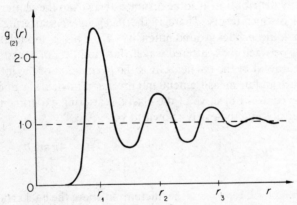

Figure 1.4 Typical radial distribution function

origin is located on a representative central particle passes through successive molecular shells beyond the central region from which neighbouring particles are geometrically excluded. The fluctuation about unity represents the probability, relative to a random distribution for which $g_{(2)}(r) = 1\cdot00$, of simultaneously finding a second particle at a distance r from the first, given that there is one at the origin. Higher-order distribution functions are not

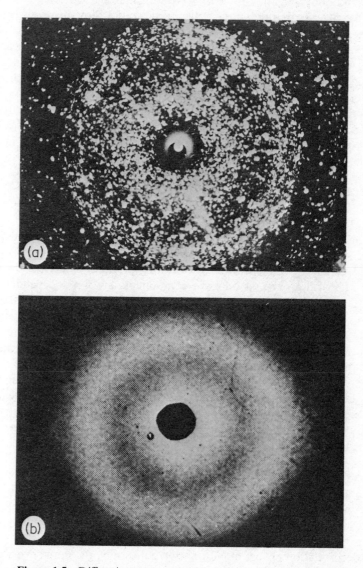

Figure 1.5 Diffraction patterns of (a) polycrystalline ice and (b) liquid water

experimentally accessible, although indirect data on the triplet distribution $g_{(3)}$ has recently become available. Fortunately for the pair theory of liquids it is the two-particle distribution $g_{(2)}(r)$ which plays the central role.

The wavelength of X-radiation and the de Broglie wavelength of fast neutrons are of the order of the interparticle spacing, and so those scattering processes are of particular importance in structural determinations in condensed matter. The velocity of propagation of these waves through the scattering assembly is so fast that the distribution appears effectively 'static'. Slow-neutron scattering will, however, detect the motion of the particles in the course of traversing the specimen, and the de Broglie waves are effectively 'Döppler shifted' in scattering from the moving scatterers. By analysing the energy distribution of the scattered neutrons about the elastic or beam energy, the nature of the kinetic processes operating in the liquid may be determined.

Optical and electron scattering has also been used under certain circumstances —the former in critical scattering studies, and the latter in reflection electron diffraction investigations of the surface structure of liquid metals and in transmission studies of gases.

Molecular Motion in Liquids

Computer simulations of the dynamic behaviour of assemblies of interacting particles show that the molecular trajectories consist of diffusive motions of temporal extent τ_D interspersed by vibratory motions of period τ_V, where $\tau_D \gg \tau_V$. The vibratory and diffusive components are not independent—the low-frequency vibratory modes merge into the diffusive contributions. Nevertheless, two distinct time scales may be discerned, and whilst τ_V remains virtually the same either side of the melting point, the period for a diffusive motion τ_D increases dramatically upon solidification, reflecting the very low coefficients of self-diffusion ($\sim 10^{-9}$ cm^2 s^{-1}) in the solid.

There is dynamical correlation in the molecular motion in a dense liquid—a representative particle may be regarded as being 'caged' by the force field of its nearest neighbours for a time $\sim \tau_D$ after which the environment has relaxed and the motion of the representative particle can no longer be regarded as correlated with its initial dynamical evolution. This dynamical correlation is expressed in terms of the *velocity autocorrelation function* whose Fourier transform, the power spectrum, gives the distribution of energy amongst the diffusive and vibratory modes operating in the liquid. These functions may be directly investigated by slow-neutron-diffraction techniques and confirm the qualitative subdivision of the molecular motion into diffusive and vibratory components.

This decorrelation or loss of information concerning the initial evolutionary history fixes a finite 'memory' to the evolutionary process and is taken to underlie the essential irreversibility operating in interacting molecular assemblies. The irreversible trend towards equilibrium appears at first sight inconsistent with the mechanically reversible equations of molecular motion: why does the evolution always progress towards equilibrium and why can't an initial non-equilibrium state spontaneously recur? In this form the problem is known as Loschmidt's Reversibility Paradox. In another form, almost by way of an answer, it appears as Zermelo's Recurrence Paradox based on a theorem of Poincaré according to which a system of particles of finite total volume and energy and subject to forces which depend only on their spatial coordinates any given initial state must, in general, recur to any state of specified accuracy infinitely often. So this tells us that initial states *may* be reversibly reattained. Associated with all this is the axiomatic law of increase of entropy. The reconciliation of these aspects is by no means complete, although we have a reasonable understanding of the irreversible trend towards equilibrium in dilute gases. At liquid densities, however, we have only a formal understanding of the approach to equilibrium. These aspects are discussed in some detail in Chapters 11 and 12.

Liquid Mixtures

We are able to make some quantitative and semi-quantitative progress in the discussion of a liquid mixture AB by first of all taking account of AA and BB interactions on the basis of the single-component theory outlined in this book. Then the cross-phenomena AB may be incorporated in terms of *effective* molecular parameters

$$\sigma_{AB} = \frac{\sigma_A + \sigma_B}{2}, \qquad \varepsilon_{AB} = \sqrt{(\varepsilon_A \varepsilon_B)} \tag{1.11}$$

for the molecular diameter and the well depth (Figure 1.1). The latter expression accounts, albeit crudely, for particle polarizability which will undoubtedly occur when two distinct species are brought together. In this way the properties of the mixture are described as a combination of AA, BB and AB contributions, and the final result is compounded as a simple sum of the separate components.

Obviously this is only an approximate approach to binary liquid systems. Nevertheless, it has yielded a number of very useful results, and although we shall not be concerned with liquid mixtures in this book, we shall briefly discuss liquid-alloy systems in Chapter 8 where this linear combination approach seems to work quite successfully.

References

J. A. Pryde, *The Liquid State*, Hutchinson University Library (1966). An elementary and readable account of the fundamentals of liquid state physics.
P. A. Egelstaff, *An Introduction to the Liquid State*, Academic Press (1967). A general introduction to liquid state physics with a pronounced emphasis on structural and dynamic investigations by scattering techniques.
J. S. Rowlinson, *Liquids and Liquid Mixtures*, Butterworths (1969). Discussions of non-simple molecular systems, including mixtures, mainly in thermodynamic terms.

CHAPTER 2

Theory of Imperfect Gases

Introduction

With decreasing temperature and increasing density the departure from idealism of a real gas is, of course, to be attributed to a development of interatomic coupling negligible at lower densities and higher temperatures. In the high-temperature approximation, violent, binary impulsive encounters are imagined to account for the principal kinetic features of the gas, whilst in the intercollision period the molecules execute free trajectories terminating in a further impulsive encounter. In addition to the assumption of no spatial correlation of the particles, or a random configuration of atomic centres, the hypothesis of *molecular chaos* is generally invoked, which excludes any dynamical correlation before and after collision: these assumptions in themselves cannot remain tenable at higher densities, and we must anticipate both spatial and dynamical correlation— correlation in phase—in a real system at anything other than the lowest densities and highest temperatures.

The thermodynamic functions of the system will, of course, depend both upon the spatial and dynamical distribution of the particles in phase, and the specification of the phase distribution is generally a very difficult problem indeed. Only under certain drastic assumptions regarding the spatial distribution, such as a random configurational distribution, can any immediate progress be made. Otherwise, the development incorporating the processes of spatial correlation becomes rapidly intractable. Instead of a direct attack on the many-body problem, the departure from idealism is approached in terms of a *cluster expansion*, in fact the virial series, in which the terms in the virial expansion account for the effects of coupling within the assembly amongst clusters of two, three, four, ... particles. As we shall see, the computational labour involved in evaluating the higher virial coefficients in practice restricts the virial description to no more than five or six terms. Indeed, Boltzmann in 1899, using some calculations of von Laar, was able to write

$$P + a\rho^2 = kT\rho(1 + b\rho + 0.625b^2\rho^2 + 0.2869b^3\rho^3 + \ldots) \qquad (2.1)$$

($b = 2\pi\sigma^3/3$, σ = atomic diameter) and regard it as a more precise form of van der Waals' equation. With all the indispensible advantages of modern statistical mechanical analysis and fast electronic computers we have been able to add only three more (approximate) terms

$$+0.1103b^4\rho^4 + 0.0386b^5\rho^5 + 0.0127b^6\rho^6 \qquad (2.2)$$

The situation is even less encouraging for more realistic interactions.

14

It was believed at one time that the virial series should, in principle, provide a satisfactory description of the thermodynamic features of the strongly coupled liquid phase, provided enough virial coefficients could be evaluated. Unfortunately, it appears that the radius of convergence of the virial series is such that the expansion will fail well below liquid densities, and that there can be no hope that it will ever yield an adequate description of the liquid phase. The study of low-density systems has considerable methodological interest, however, and when we come to consider high-density liquid systems we shall encounter many of the same formal difficulties. The emphasis at higher densities is on a configurational or structural description of the fluid, in addition to the thermodynamic functions which are, of course, intimately related to the structure. Structural considerations are of secondary interest in dilute systems however, and the analytic effort is rather directed at the evaluation of the N-body partition function Z_N which, as we shall shortly demonstrate, provides direct access to all the principal thermodynamic functions of state.

The Canonical Form

The most complete specification of the phase distribution of an assembly of N identical, coupled structureless particles is the N-body distribution function $f_{(N)}(\mathbf{p}_1, \ldots, \mathbf{p}_N, \mathbf{q}_1, \ldots, \mathbf{q}_N)$ where $\mathbf{p}_1, \ldots, \mathbf{p}_N, \mathbf{q}_1, \ldots, \mathbf{q}_N$ (generally written $\mathbf{p}^N, \mathbf{q}^N$ without ambiguity) represent the canonically conjugate momentum and spatial variables of the N-body distribution. For an equilibrium, time-independent distribution, $f_{(N)}$ may be related to the system Hamiltonian through the canonical form

$$f_{(N)}(\mathbf{p}^N, \mathbf{q}^N) = \frac{1}{Z_N} \exp\left\{ -\frac{\mathscr{H}(\mathbf{p}^N, \mathbf{q}^N)}{kT} \right\} \qquad (2.3)$$

where Z_N is a constant of proportionality termed the N-body phase partition function, or *zustandsumme*. Z_N, as we shall see, plays a central role in the description of the thermodynamic functions of state of a classical system, and, indeed, of a quantal system provided we identify Z_N as the *Slater sum*. The system Hamiltonian may for many, but not all, classical systems be split up into its spatial and dynamical components:

$$\mathscr{H}(\mathbf{p}^N, \mathbf{q}^N) = \sum_{i=1}^{N} \frac{\mathbf{p}_i^2}{2m_i} + \Phi_N(\mathbf{q}_1, \ldots, \mathbf{q}_N) \qquad (2.4)$$

and fortunately we shall be concerned here only with such separable Hamiltonians. m_i represents the mass of particle i, of momentum $\mathbf{p}_i$, and $\Phi_N(\mathbf{q}_1, \ldots, \mathbf{q}_N)$ represents the total configurational energy which depends explicitly upon the unknown spatial distribution of particles $(\mathbf{q}_1, \ldots, \mathbf{q}_N)$. Before elaborating on equation (2.4) whose configurational component evidently underlies the departure of a real gas from idealism, we shall first examine some of the formal aspects of the distribution $f_{(N)}$, and its relation to the phase partition function Z_N.

Since there is an inherent limit to the precision with which we are able to specify the phase $\mathbf{p}_i$, $\mathbf{q}_i$ of the ith particle, we regard phase space as having a granularity whose unit of three-dimensional phase volume is h^3. The quantity

$$h^{-3N} f_{(N)}(\mathbf{p}_1, \ldots, \mathbf{p}_N, \mathbf{q}_1, \ldots, \mathbf{q}_N)\, d\mathbf{p}_1, \ldots, d\mathbf{p}_N, d\mathbf{q}_1, \ldots, d\mathbf{q}_N \qquad (2.5)$$

then represents the probability of finding particle 1 within the volume element $d\mathbf{p}_1\, d\mathbf{q}_1$ located at $\mathbf{p}_1$, $\mathbf{q}_1$, whilst particle 2 is *simultaneously* located within $d\mathbf{p}_2\, d\mathbf{q}_2$ of $\mathbf{p}_2, \mathbf{q}_2$, and so on, up to particle N simultaneously located within $d\mathbf{p}_N\, d\mathbf{q}_N$ of $\mathbf{p}_N\mathbf{q}_N$. Of course, we have accounted here only for one possible *complexion* of the phase distribution: we have neglected the fundamental indistinguishability of the particles, and any permutation of the N particles amongst the N possible phase locations represents an equally acceptable representation of the distribution in phase. We therefore have to divide the enormous number of complexions which would arise, were the particles distinguishable, by $N!$ to account for their indistinguishability. We shall work in the *quasi-classical approximation* throughout, except when dealing with specifically wave-mechanical situations, and this differs from the classical Boltzmann expression by the factor $(N!)^{-1}$. Whilst we may be working in the temperature and density range where classical mechanics yield a very good approximation, we cannot go so far as to neglect the essential indistinguishability of the particles. $\ln Z_N$ in the classical and quasi-classical approximations differs by the additive constant $\ln N!$, and certain quantities, such as the pressure and energy, arise as *derivatives* of $\ln Z_N$ and hence remain unaffected. The entropy and free energy will differ, however, and the problem is identified as the *Gibbs paradox*. Now, if we were to integrate (2.5) over all possible values of the spatial and dynamical coordinates, the certainty that the N particles are located somewhere within the phase volume enables us to write the normalization condition,

$$1 = \frac{1}{N! h^{3N}} \int \ldots \int f_{(N)}(\mathbf{p}^N, \mathbf{q}^N)\, d\mathbf{p}^N\, d\mathbf{q}^N$$

which, from equations (2.3) and (2.4) may be written

$$Z_N = \frac{1}{N! h^{3N}} \int \ldots \int \exp\left\{ -\sum_{i=1}^{N} \mathbf{p}_i^2 / 2 m_i k T \right\} d\mathbf{p}_1 \ldots d\mathbf{p}_N$$

$$\times \int \ldots \int \exp\left\{ -\Phi_N(\mathbf{q}_1, \ldots, \mathbf{q}_N)/kT \right\} d\mathbf{q}_1 \ldots d\mathbf{q}_N \qquad (2.6)$$

i.e. $Z_N = Z_P Z_Q$, where

$$Z_P = \frac{1}{h^{3N}} \int \ldots \int \exp\left\{ -\sum_{i=1}^{N} \mathbf{p}_i^2 / 2 m_i k T \right\} d\mathbf{p}_1 \ldots d\mathbf{p}_N \qquad (2.7)$$

$$Z_Q = \frac{1}{N!} \int \ldots \int \exp\left\{ -\Phi_N(\mathbf{q}_1, \ldots, \mathbf{q}_N)/kT \right\} d\mathbf{q}_1 \ldots d\mathbf{q}_N \qquad (2.8)$$

This factorization of the total phase partition function into a momentum (Z_P)

and configurational (Z_Q) component has arisen directly as a consequence of the separability of the Hamiltonian (2.4). We see, incidentally, the basis for the German expression *zustandsumme* (= sum-over-states) from equations (2.6), (2.7) and (2.8). The momentum partition Z_P may be further factorized into a product of single-particle components since

$$\exp\left\{-\sum_{i=1}^{N} \mathbf{p}_i^2/2m_ikT\right\} = \prod_{i=1}^{N} \exp\left\{-\mathbf{p}_i^2/2m_ikT\right\} \tag{2.9}$$

Moreover, we should emphasize that Z_P develops *independently* of the configuration $(\mathbf{q}_1,\ldots,\mathbf{q}_N)$, and to this extent describes the equilibrium momentum distribution of the assembly, irrespective of state, solid, liquid or gas. Evaluation of the integral Z_P is readily achieved, yielding

$$Z_P = \left(\frac{2\pi mkT}{h^2}\right)^{3N/2} \tag{2.10}$$

and has dimensions (volume)$^{-N}$. The configurational integral Z_Q is incomparably more difficult to evaluate, and may be said to represent the fundamental problem in the theory of fluids. It is to the evaluation of this function that the principal research effort has been directed and it underlies the thermodynamic and configurational description of fluids, particularly at liquid densities.

At very low densities we may, as a first approximation, propose that the particles are sufficiently weakly coupled that the total configurational energy $\Phi_N(\mathbf{q}_1,\ldots,\mathbf{q}_N)$ is zero: the entire internal energy in this case being kinetic. This being so, the configurational partition function becomes (cf. equation (2.8))

$$Z_Q = \frac{V^N}{N!}$$

whereupon the total partition function in this case is

$$Z_N = \left(\frac{2\pi mkT}{h^2}\right)^{3N/2} \frac{V^N}{N!} \tag{2.11}$$

The equation of state can be readily obtained from the total phase partition function Z_N by the general relation

$$P = \frac{1}{\beta}\left(\frac{\partial \ln Z_N}{\partial V}\right)_T \quad \text{where } \beta = (kT)^{-1} \tag{2.12a}$$

as can the internal energy,

$$U = kT^2 \frac{\partial(\ln Z_N)_V}{\partial T} \tag{2.12b}$$

These are standard statistical mechanical results, and we direct the reader to any standard statistical mechanics text for their derivation, together with the other thermodynamic relations to Z_N. From (2.11) we have

$$\ln Z_N = \frac{3N}{2} \ln \left(\frac{2\pi mkT}{h^2}\right) + N \ln V - \ln N! \tag{2.13}$$

(the quasi-classical approximation differs from the classical Boltzmann expression by the additive constant $-\ln N!$) whereupon, from (2.12)

$$P = \frac{NkT}{V} \tag{2.14}$$

which is, of course, the ideal-gas equation of state. The momentum distribution is Maxwell–Boltzmann whilst the configuration is assumed entirely random. In other words, this model takes no account of either the long-range attractive forces, nor of the short-range repulsive core. Van der Waals provided the first analytic excursion from idealism in attempting to account for the finite size of the molecular core, and for the long-range attractive component of the intermolecular pair potential (Figure 2.1).

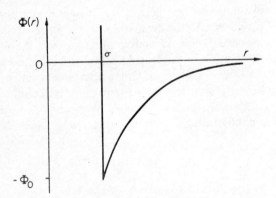

Figure 2.1   The van der Waals interaction showing rigid repulsive core of collision diameter σ, and long-range attractive branch

We may extend the above discussion to take account of the principal features of the atomic interaction as follows. If we assume that a given particle in the assembly moves in the effective field of the remaining $(N-1)$ particles, and that, moreover, this particle does not affect the distribution of the remainder, then the assembly may be treated as a system of N *independent* particles, each moving in some effective potential $\Phi_{\text{eff}}(\mathbf{q})$. The momentum integral remains the same as for a classical gas (2.10), whilst the total partition becomes

$$Z_N = \frac{1}{N!} \left(\frac{2\pi mkT}{h^2}\right)^{3N/2} \left[\int \exp\left(-\Phi_{\text{eff}}(\mathbf{q})/kT\right) d\mathbf{q}\right]^N \tag{2.15}$$

where the configurational integral ranges over the entire volume of the container, V. If we assume, with van der Waals, a rigid molecular core of the form shown in Figure 2.1, then in these regions $\Phi_{\text{eff}}(\mathbf{q}) \to \infty$: the integrand vanishes in this excluded volume V_x. Over the remaining accessible region $(V - V_x)$ we make the further assumption that the potential adopts some *mean* value $\overline{\Phi}_{\text{eff}}$, in which case equation (2.15) becomes

$$Z_N = \frac{1}{N!} \left(\frac{2\pi m k T}{h^2} \right)^{3N/2} [(V - V_x) \exp(-\overline{\Phi}_{\text{eff}}/kT)]^N \qquad (2.16)$$

It remains to identify the quantities V_x and $\overline{\Phi}_{\text{eff}}$: then we may establish the equation of state from the relation (2.12). The mean effective potential $\overline{\Phi}_{\text{eff}}$ acting within the assembly may be related to the mean potential developed between a pair $\overline{\Phi}$ as

$$\overline{\Phi}_{\text{eff}} = \tfrac{1}{2} N \overline{\Phi} \qquad (2.17)$$

where the factor $\frac{1}{2}N$ represents the number of *pairs* of molecules. For the functional form of the potential shown in Figure 2.1 we may assume

$$\begin{aligned} \Phi(r) &= +\infty & r < \sigma \\ &= -\left(\frac{\sigma}{r} \right)^s \Phi_0 & r \geqslant \sigma \end{aligned} \right\} \qquad (2.18)$$

so that the mean potential per unit volume developed between a pair of molecules, assuming a uniform (random) distribution of particles beyond the collision diameter, is

$$\overline{\Phi} = \frac{1}{V} \int \Phi(\mathbf{q}) \, d\mathbf{q} = -\frac{4\pi \Phi_0}{V} \int_\sigma^r \Phi(r) r^2 \, dr$$

$$= -\frac{4\pi \Phi_0}{V} \int_\sigma^r \left(\frac{\sigma}{r} \right)^s r^2 \, dr \qquad (2.19)$$

For the integral to converge, we must assume $s > 3$. This condition is generally satisfied for electrically neutral particles for which $s \sim 6$. We then have for the mean effective potential (2.17)

$$\overline{\Phi}_{\text{eff}} = -\frac{2\pi \sigma^3}{3} \left(\frac{3}{s-3} \right) \Phi_0 \frac{N}{V} = -a\frac{N}{V} \qquad (2.20)$$

V_x was taken to represent the volume excluded per particle, and so we have (see Figure 2.2)

$$V_x = \frac{2\pi}{3} \sigma^3 N = bN \qquad (2.21)$$

It follows directly from equations (2.12) and (2.16) that

$$P = \frac{1}{\beta} \left(\frac{\partial \ln Z_N}{\partial V} \right)_T = \frac{1}{\beta} \frac{\partial}{\partial V} [N \ln (V - V_x) - N\beta \overline{\Phi}_{\text{eff}}] \qquad (2.22)$$

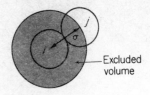

Excluded volume

Figure 2.2 Geometrical exclusion between rigid spheres. The per particle excluded volume $b = 2\pi\sigma^3/3$

i.e.

$$\left(P + \frac{aN^2}{V^2}\right)\left(\frac{V}{N} - b\right) = kT, \qquad \frac{N}{V} = \rho \qquad (2.23)$$

where a and b are defined above, and generally $s \sim 6$, presenting no difficulties of convergence. Equation (2.23) is, of course, van der Waals equation and may be readily rewritten in virial form:

$$P + a\rho^2 = kT\rho(1 + b\rho + b^2\rho^2 + b^3\rho^3 + \ldots) \qquad (2.24)$$

which is to be compared to the Boltzmann equation of state (2.1). The attractive component of the pair potential affects only the parameter a, whilst all the features of geometric exclusion are contained in the right-hand sides of the virial expansions through the coefficients of the density terms. The discrepancy between the Boltzmann–von Laar equation and that of van der Waals arises from the incomplete assessment of exclusion in the van der Waals case. This may be attributed to the complicated geometrical gaps which arise in the packing of spheres—gaps deemed accessible, which of course they are not, in the van der Waals treatments. As we might expect, the two equations of state show a progressive discrepancy with increasing density. The situation becomes much more complicated for more realistic interactions, and we shall be forced to abandon this simple geometric interpretation. Nevertheless, in establishing the equation of state (2.23), no specific appeal at low-density or high-temperature approximations was made, and we might therefore anticipate that (2.23) should be capable of describing some of the general features of the equation of state for dense liquids as well as dilute gases.

The Grand Canonical Form

We have in the previous section related the N-body phase distribution to spatial and dynamical features of the assembly through the Hamiltonian of the system, expressed in canonical form. In this way we were able to arrive at an expression for the phase partition function. Suppose now we have to deal with an *open* mechanical system in which the number of particles is not fixed. In this case the system Hamiltonian becomes an explicit function of the number N of particles and we may now write, in grand canonical form,

$$\mathscr{F}_{(N)}(\mathbf{p}^N, \mathbf{q}^N) = \frac{1}{\mathscr{Z}_N} \exp\left\{\frac{\mu N - \mathscr{H}(\mathbf{p}^N, \mathbf{q}^N)}{kT}\right\} \qquad (2.25)$$

where $\mathscr{Z}_N$ is termed the *grand partition function* and has the same normalizing role as Z_N in (2.3). If μ is identified with the macroscopic chemical potential then the thermodynamic functions of the system are obtained.

In evaluating $\mathscr{Z}_N$ from the normalization condition which applies to the distribution $\mathscr{F}_{(N)}(\mathbf{p}^N, \mathbf{q}^N)$, we must sum over all possible particle numbers of the average over the phase:

$$\mathscr{Z}_N(V, T) = \sum_{N \geqslant 0} \exp\left\{ \frac{\mu N}{kT} \right\} Z_N \qquad (2.26)$$

Obviously, if the number of particles is constant $\mu = 0$ and (2.25) reduces to canonical form. The importance of the grand canonical representation lies in its application to open mechanical systems, and it is obviously of more general application than the closed canonical form. We shall not, however, be concerned with grand canonical functions in this book, and we merely point out here that the thermodynamic functions appropriate to an open system are related through $\mathscr{Z}_N$ in an analogous way to their canonical counterparts, and we direct the reader to one of the standard statistical mechanical texts for further details.

The general evaluation of the equation of state for a coupled assembly of particles in terms of the phase partition function Z_N is a formidable problem, and little progress other than in a formal sense is possible without approximation. The difficulty arises as usual in the evaluation of the configurational partition function

$$Z_Q = \frac{1}{N!} \int \ldots \int \exp\left\{ -\Phi_N(\mathbf{q}_1, \ldots, \mathbf{q}_N)/kT \right\} d\mathbf{q}_1, \ldots, d\mathbf{q}_N \qquad (2.27)$$

for what do we write for $\Phi_N(\mathbf{q}_1, \ldots, \mathbf{q}_N)$? A knowledge of $\Phi_N(\mathbf{q}_1, \ldots, \mathbf{q}_N)$ does, of course, presuppose a knowledge of the molecular configurational distribution $(\mathbf{q}_1, \ldots, \mathbf{q}_N)$ and this is quite unknown. Some progress may be made by assuming that the potential is pairwise additive, that is, the total potential $\Phi_N(\mathbf{q}_1, \ldots, \mathbf{q}_N)$ may be represented by the sum of potentials developed between *pairs* of molecules, care being taken not to count the interactions twice. Thus, in the pair approximation,

$$\Phi_N(\mathbf{q}_1, \ldots, \mathbf{q}_N) = \sum_{i>j}^{N} \sum^{N} \Phi(ij) \qquad (2.28)$$

This approximation underlies the entire modern statistical theories of fluids, primarily for the considerable simplification which ensues. It should be emphasized, however, that the implicit assumption in the approximation is that the interaction between a pair of molecules remains unmodified in the presence of a third. There is now a body of experimental and theoretical evidence which suggests that *three-body* contributions are significant, and we shall later consider triplet effects in some detail. For the purposes of analytical discussion, however, we shall continue in the pair approximation.

22

First, however, we introduce the Mayer f-function—a function which proves of central importance in the theory of fluids. The f-function is simply defined as

$$f(r_{ij}) = \exp\left\{ -\frac{\Phi(r_{ij})}{kT} \right\} - 1 \qquad (2.29)$$

and its relation to the pair potential is shown in Figure 2.3. As we see, the f-function remains bounded within the core, even though $\Phi(ij) \to \infty$; indeed, the

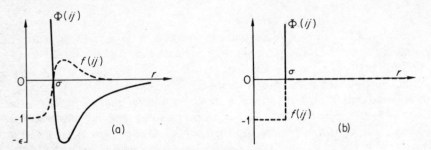

Figure 2.3 (a) The Mayer f-function (broken curve) associated with the realistic interaction $\Phi(ij)$:

$$f = \left[\exp\left(-\frac{\Phi(ij)}{kT} \right) - 1 \right]$$

In this case f is temperature-dependent. (b) The Mayer f-function (broken curve) and the hard-sphere interaction. In this case f is seen to be a negative step function. Generally the f-function remains bounded even though the core potential becomes infinite

f-function is particularly simple in the case of the hard-sphere potential. The Mayer function is, it should be remembered, temperature dependent (see equation (2.29)) whilst $\Phi(ij)$ in general is not. Indeed, as the temperature of the assembly is increased the effects of the repulsive core dominate the interaction and the positive region of the f-function shown in Figure 2.3(a) becomes washed out, tending to the negative step function shown in Figure 2.3(b).

We may take advantage of the pair potential expression of Φ_N given in (2.28). Under these circumstances

$$\exp\left\{ -\frac{\Phi_N}{kT} \right\} = \prod_{i>j}^{N}\prod^{N} \exp\left\{ -\frac{\Phi(r_{ij})}{kT} \right\} \qquad (2.30)$$

From equation (2.27) the N-body configurational partition function may be written

$$Z_Q = \frac{1}{N!} \int \dots \int \prod_{i>j}\prod (1 + f_{ij})\, d\mathbf{1} \dots d\mathbf{N} \qquad (2.31)$$

where we have used $\int \dots \int d\mathbf{1} \dots d\mathbf{N}$ to signify integration over all positions of molecules $1 \ldots N$. Expansion of the integrand yields

$$\prod_{N \geqslant i > j \geqslant 1} \prod (1 + f_{ij}) = 1 + \sum_i \sum_j f_{ij} + \sum_i \sum_j \sum_k \sum_l f_{ij} f_{kl} + \cdots$$

$$+ \sum \cdots \sum \cdots \sum f_{ij} f_{kl} \cdots f_{mn} + \cdots \quad (2.32)$$

the terms of which can be represented in diagrammatic form. The diagrams which arise can be split up and reassembled into *irreducible clusters*: these are defined as mappings or graphs in which each point or molecule is connected by an f-bond to at least two other points. Thus:

$$\left. \begin{array}{l} \{\ \text{o—o}\ \} \ \text{2-body clusters between molecules } i, j \\[2mm] \{\ \triangle\ \} \ \text{3-body clusters between molecules } i, j, k \\[2mm] \{\ \square, \boxtimes, \boxtimes\ \} \ \text{4-body clusters between molecules } i, j, k, l \end{array} \right\} \quad (2.33)$$

where the exceptional case of two interacting molecules is included as the lowest irreducible cluster. The general product in this series consists of the sum over all connected products of the same order, n. By this is meant the sum over all n-body clusters which are topologically distinct. The number of such clusters in each n-body subset increases rapidly with n. Thus, as we see from (2.33), there is one 2-body cluster, one 3-body cluster and three 4-body clusters. Beyond this there are ten 5-body clusters and fifty-six 6-body clusters. Each cluster within a given subset is weighted according to the frequency of its occurrence in the expansion (2.32). The combinatorial analysis necessary to establish the general expansion (2.32) in terms of the correctly weighted n-body subsets (2.33) is formidable, but is discussed in the literature. Further, indistinguishability of the particles within each cluster means that permutation of the particles amongst the sites does not produce a topologically distinct diagram. We must therefore divide the weighting of each diagram by $n!$. In the evaluation of the partition function we must only sum over physically distinct states, otherwise we shall be confronted with a discrepancy between the predicted and observed free energy and entropy—the Gibbs paradox.

It follows from the preceding discussion that the cluster expansion of the *total* phase partition function is

$$Z_N = \left(\frac{2\pi mkT}{h^2} \right)^{3N/2} \int \cdots \int \left\{ 1 + \frac{1}{2!} \frac{N!}{(N-2)!} \text{ o—o} + \frac{1}{3!} \frac{N!}{(N-3)!} \triangle \right.$$

$$\left. + \frac{1}{4!} \frac{1}{(N-4)!} [3\ \square + 6\ \boxtimes + \boxtimes] + \cdots \right\} d\mathbf{1} \ldots d\mathbf{N} \quad (2.34)$$

where the coefficient of the n-body subset, $N!/n!(N - n)!$ represents the number of ways the n-body cluster can be formed from the complete set of N molecules. Indistinguishability is being taken into account termwise in (2.34): the weighting factors first appear in the four-body diagrams. Integration yields

$$Z_N = \left(\frac{2\pi mkT}{h^2}\right)^{3N/2} \left\{ V^N + \frac{N}{2!}(N-1)V^{N-1} \bullet\!\!-\!\!\bullet \right.$$

$$\left. + \frac{N}{3!}(N-1)(N-2)V^{N-2} \triangle + \dots \right\} \qquad (2.35)$$

where the *cluster integrals* are defined as follows.

$$\left. \begin{array}{l} \bullet\!\!-\!\!\bullet \\ \scriptstyle 1 \;\; 2 \end{array} \equiv \int \circ\!\!-\!\!\circ \, d2 = \int f_{12} \, d2 \right.$$

$$\left. \begin{array}{l} \triangle \\ \scriptstyle 1 \;\; 2 \end{array} \equiv \iint \triangle \, d2 \, d3 = \iint f_{12} f_{23} f_{31} \, d2 \, d3 \right\} \qquad (2.36)$$

$$\left. \begin{array}{l} \boxtimes \\ \scriptstyle 1 \;\; 2 \end{array} \equiv \iiint \boxtimes \, d2 \, d3 \, d4 = \iiint f_{12} f_{13} f_{34} f_{14} f_{23} \, d2 \, d3 \, d4 \right.$$

etc.

The cluster integrals are evaluated by a Monte Carlo technique whereby trial configurations of the *n*-body cluster are set up on the computer, and the configuration is tested to see whether the range and value of the bonds is consistent with the form of the integrand. If the configuration is inconsistent it is rejected. If it is consistent the configuration is accepted and contributes to the cluster integral. If accurate estimates are to be obtained it is evident that the trial configurations must be sufficiently 'fine-grained', and in consequence many hundreds of thousands of trial configurations must be set up. Obviously, for heavily bonded many-body clusters the likelihood of a 'successful' configuration is correspondingly diminished, and if the estimates are not to be subject to a large statistical error arising from the small number of successful configurations, it is clear that the Monte Carlo evaluation of the diagram will be a very un-economical process indeed.

A simple construction helps in the evaluation of the cluster integrals: the common volume associated with two spheres of unit *radius* may be shown to be *numerically* identical to the two-body cluster integral $\int f_{12} \, d2$ for hard spheres of unit *diameter*. Similarly for the three-body cluster integral $\iint f_{12} f_{23} f_{31} \, d2 \, d3$. The net sign varies of course according as the number of f-bonds is odd or even, since the value of the hard-sphere f-bond is either -1 or zero. The construction may be extended to realistic interactions provided we remember the long-range form of the f-function is positive and the entire function is temperature-dependent (Figure 2.4).

We have now established a cluster expansion of the N-body phase partition function. All the principal thermodynamic quantities now follow. The equation of state, for example

$$P = \frac{1}{\beta}\left(\frac{\partial \ln Z_N}{\partial V}\right)_T, \qquad \beta = (kT)^{-1}. \qquad (2.37)$$

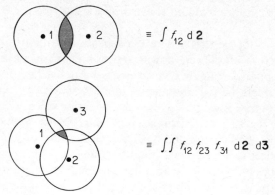

Figure 2.4 Geometry for the evaluation of the cluster integrals •-•, $\triangle$

From equation (2.35) we have

$$\ln Z_N = N \ln V + \ln \left\{ 1 + \frac{N(N-1)}{2!V} \bullet\!\!-\!\!\bullet + \frac{N(N-1)(N-2)}{3!V^2} \triangle + \ldots \right\}$$
$$+ \frac{3N}{2} \ln \left(\frac{2\pi mkT}{h_0^2} \right) \tag{2.38}$$

Now, since N is very large, $N(N-1) \sim N^2$, etc. and for $x \ll 1$, $\ln(1+x) \sim x$ so that from (2.37) and (2.38) the equation of state immediately follows:

$$P = \frac{NkT}{V} \left\{ 1 - \frac{N}{V} \frac{1}{2!} \bullet\!\!-\!\!\bullet - \left(\frac{N}{V} \right)^2 \frac{2}{3!} \triangle - \ldots \right\} \tag{2.39}$$

which is, of course, the virial expansion

$$P = \frac{NkT}{V} \left\{ 1 + \frac{N}{V} B_2(T) + \frac{N^2}{V^2} B_3(T) + \ldots \right\} \tag{2.40}$$

where $B_2(T)$, $B_3(T)$, ... are the virial coefficients.

We are now able to understand the physical significance of the cluster integrals and their identification in the second and subsequent terms in the virial expansion. Departure from idealism is accounted for in terms of the development of clusters of two, three, ... particles and so on through successively larger groupings in simultaneous interaction. Individual diagrams have little immediate physical significance, but taken collectively as a group *of the same order*, they have the significance which has just been described. Obviously, the development of many-body clusters will depend essentially on the density of the system, and at low densities we might anticipate that the effects of association amongst pairs and triplets of particles will provide an adequate description of the initial departure from idealism. This is not to say, incidentally, that the molecules remain in permanent association: the clusters are continuously forming and collisionally disrupting and indeed above the critical temperature cannot form a bound assembly.

The convergence of the virial expansion depends essentially upon the density being small. There is, it has been shown, no possibility of the series providing an adequate description for highly connected assemblies at liquid densities when $\sim 10^2$ particles might be in simultaneous interaction. On the cluster-diagrammatic interpretation this would involve the evaluation of an enormous number of impossibly difficult cluster integrals, even if the expansion could be extended to liquid densities in principle, and it is quite clear that the virial series cannot account for collective phenomena involving the cooperative behaviour of a very large number of molecules within the fluid, such as phase transitions.

The cluster expansion of the internal energy at constant volume follows directly from equations (2.12b) and (2.38):

$$U = kT^2 \frac{\partial(\ln Z_N)}{\partial T}$$

We see that the equipartition value $3NkT/2$ is supplemented by a cluster series as follows:

$$\frac{U}{V} = \frac{3}{2}\rho kT\left[1 + \frac{2}{3}\rho T\frac{\partial B_2}{\partial T} + \frac{2}{3}\rho^2 T\frac{\partial B_3}{\partial T} + \cdots \right] \qquad (2.41)$$

A test of the internal consistency of the theory now follows since the coefficients of the density expansion of the internal energy of the assembly should be simply related to the temperature derivative of the density coefficients of the equation of state at constant temperature (equation (2.40)).

We should finally make the observation that the results obtained so far are exact within the pair approximation. The evaluation of the cluster integrals therefore provides the first few virial coefficients exactly: these results will become of particular importance later on when we test our approximate theories developed for use at liquid densities. We might expect these theories to yield reasonably accurate values of the virial coefficients, and such a comparison should provide a useful test of the confidence which can be placed in the approximations.

The Cluster Expansion of Ree and Hoover

The difficulty in evaluating the cluster integrals in practice restricts the virial equation of state to the first few terms: the computational labour in evaluating the rapidly diverging number of diagrams which develop with increasing cluster order makes a direct evaluation prohibitive. The number of cluster integrals contributing to B_n is $\sim 2^{\frac{1}{2}n(n-1)}/n!$, which is over 600 for $n = 8$. Ree and Hoover have proposed a very elegant means of computation which effects a drastic reduction in the number of diagrams which have to be evaluated.

We continue in the pair approximation, and begin by defining a Ree–Hoover function,

$$\tilde{f}(r_{ij}) = \exp\left(-\Phi(r_{ij})/kT\right) \tag{2.42}$$

which we denote by a wiggly line $\sim$. The Ree–Hoover function has the following relation to the Mayer f-function:

$$1 = \tilde{f}_{ij} - f_{ij} \tag{2.43}$$

and we denote the function $(\tilde{f}_{ij} - f_{ij})$ by a broken line $------$. Whenever two points in one of the diagrams arising in the cluster expansion of the partition function are not connected by an f-bond, the function $(\tilde{f} - f)$ is introduced into the diagram to link the unbonded points. This of course has no *numerical* effect since the value of the bond is unity for all configurations of the cluster. It does have a dramatic effect on the number of diagrams which have to be evaluated, however. Take, for example, the irreducible cluster expansion of the virial coefficient B_4 (equation (2.34)):

$$3\;\square\!\!\!\square + 6\;\boxtimes + \boxtimes \tag{2.44}$$

and suppose we now incorporate the $(\tilde{f} - f)$ bonds between unconnected points

$$3\;\boxtimes + 6\;\boxtimes + \boxtimes \tag{2.45}$$

where, of course, the broken lines represent the $(\tilde{f} - f)$ bond. The cluster integrals may then be written:

$$3 \int \cdots \int f_{12} f_{23} f_{34} f_{14} (\tilde{f}_{13} - f_{13})(\tilde{f}_{24} - f_{24})\, \mathrm{d}1\, \mathrm{d}2\, \mathrm{d}3\, \mathrm{d}4$$

$$+ 6 \int \cdots \int f_{12} f_{23} f_{34} f_{14} f_{13} (\tilde{f}_{24} - f_{24})\, \mathrm{d}1\, \mathrm{d}2\, \mathrm{d}3\, \mathrm{d}4$$

$$+ \int \cdots \int f_{12} f_{23} f_{34} f_{14} f_{13} f_{24}\, \mathrm{d}1\, \mathrm{d}2\, \mathrm{d}3\, \mathrm{d}4 \tag{2.46}$$

If we now expand the integrals we find there is extensive cancellation, resulting in:

$$3\;\boxtimes - 2\;\boxtimes$$

where the wiggly bond represents the $\tilde{f}$-function. In the unmodified f-function formulation there are 3, 10 and 56 topologically distinct clusters which need to be evaluated for the virial coefficients B_4, B_5 and B_6, respectively. In the Ree–Hoover $\tilde{f}$-formalism the corresponding modified expressions contain, respectively, 2, 5 and 23 topologically distinct types of modified clusters. The reduction in the number of cluster integrals which have to be evaluated is considerable: their evaluation nonetheless remains formidable, and Ree and Hoover evaluated the diagrams by a Monte Carlo technique described in the previous section. For hard spheres the functions f and $\tilde{f}$ adopt a particularly simple form and the evaluation of the hard-sphere virial coefficients is correspondingly simpler. The coefficients B_2, B_3, B_4 are known exactly for a hard-sphere assembly (equation 2.1)): Ree and Hoover have determined B_5 and B_6 and the results are shown in Table 2.1. The computations may be carried out for

Table 2.1. Hard-sphere virial coefficients

B_2	$(2\pi/3)\sigma^3 = b$
B_3/b^2	0.625
B_4/b^3	0.2869
B_5/b^4	0.1103 ± 0.0003
B_6/b^5	0.0386 ± 0.0004
B_7/b^6	0.0127
B_8/b^7	0.0040

an assembly of hard discs—the two-dimensional analogue of the hard-sphere system. Here, of course, the diagrammatic configurations are confined to a *plane*, and the evaluation of the cluster integrals is correspondingly easier. The diagrams may, of course, be subjected to the same Ree–Hoover analysis. The hard-disc virial coefficients are given in Table 2.2.

Table 2.2. Hard-disc virial coefficients

B_2	$(\pi/2)\sigma^2 = b$
B_3/b^2	0.78200
B_4/b^3	0.5324 ± 0.0003
B_5/b^4	0.3338 ± 0.0005
B_6/b^5	0.1992 ± 0.0008
B_7/b^6	~ 0.115
B_8/b^7	~ 0.065

A two-dimensional gas is obviously a theoretical abstraction, and except in one or two highly contrived circumstances does not correspond to any real physical situation. Nonetheless, the role of computer simulation in fluid studies is of outstanding importance, and a two-dimensional assembly presents no difficulties from the point of view of simulation: indeed, it is rather easier to simulate a fluid in two dimensions than it is in three. The justification for the simulation of idealized assemblies of particles lies primarily in the considerable simplification of the theoretical analysis which follows the use of pair interactions of mathematical simplicity rather than physical reality, as we have seen in the Ree–Hoover development.

For hard spheres and discs the Mayer and Ree–Hoover functions are, of course, temperature independent: it therefore follows that the virial coefficients are themselves temperature-independent constants and, from equation (2.41), the internal energy in this case reduces to its equipartition value. For realistic interactions f, $\bar{f}$ and the virial coefficients are all temperature-dependent functions, and we shall consider such systems in the following sections.

Padé Approximant to the Virial Series

The Padé approximant to an infinite series enables us to estimate its asymptotic behaviour from a knowledge of the first few coefficients. The virial series

$$\sum_{i=1}^{\infty} B_{i+1}\rho^{i-1}$$

is set identical to the $P(n, m)$ Padé approximant, defined as follows:

$$\sum_{i=1}^{\infty} B_{i+1}\rho^{i-1} \equiv P(n, m) = \sum_{i=1}^{n} a_i\rho^{i-1} \Big/ \sum_{i=1}^{m} \alpha_i\rho^{i-1} \tag{2.47}$$

Expansion and comparison of coefficients on both sides of (2.47) enables us to express a_i and α_i in terms of the virial coefficients. Substitution of the known values of the coefficients B_2 through B_6 enabled Ree and Hoover to propose the equations of state (setting $a_1 = B_2$, $\alpha_1 = 1$):

$$\frac{PV}{NkT} = 1 + \frac{b\rho(1 + 0.063507b\rho + 0.017329b^2\rho^2)}{1 - 0.561493b\rho + 0.081313b^2\rho^2} \tag{2.48}$$

for hard spheres, and

$$\frac{PV}{NkT} = 1 + \frac{b\rho(1 - 0.196703b\rho + 0.006519b^2\rho^2)}{1 - 0.978703b\rho + 0.239465b^2\rho^2} \tag{2.49}$$

for hard discs. A comparison of the virial equations of state for hard spheres and hard discs (Figure 2.5) based on the coefficients up to and including B_5, up to and including B_6, and the Padé approximant shows the latter to be in excellent agreement with the computer simulation over most of the fluid branch of the isotherm. Collective phenomena based on the cooperative behaviour of very large numbers of particles cannot possibly be described in terms of five- and six-body cluster functions, whether or not they are reassembled into approximant form. We should not be surprised, therefore, that the simulated isotherm departs from the calculated curves at intermediate and high densities, in particular in the vicinity of the phase transition.

Certainly inclusion of the coefficient B_6 represents an improvement on the five-term virial both for hand spheres and hard discs. From the Padé approximant Ree and Hoover are able to estimate the virial coefficients B_7 and B_8, and obtain the results shown in Tables 2.1 and 2.2. The coefficients B_2 through B_8 are all positive, as indeed the next few presumably are if the virial is to coincide with the 'experimental' isotherm. Ree and Hoover find that for hard spheres B_{20} is the first negative coefficient, and thereafter so is approximately every sixteenth. For hard discs it appears that all coefficients are positive.

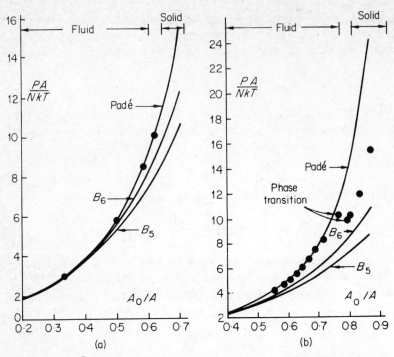

Figure 2.5 The molecular dynamics equation of state (●) for an assembly of (a) hard spheres and (b) hard discs. The virial series with coefficients up to and including B_5 and B_6 are shown, together with the Padé approximant on the basis of equations (2.48) and (2.49). None of the equations of state are able to reproduce the van der Waals loop at the solid–fluid phase transition

The Padé approximant to the virial series does not afford much physical insight into the problem. What it does provide, however, is a reasonably accurate analytic expression for the low- to intermediate-density equation of state, and to this extent proves of great convenience.

The temperature dependence of B_2 may be easily discussed in terms of the effective region of the pair potential. Thus, for close encounters arising at high temperatures the effective region of the potential curve is the positive core and the integrand $\{\exp(-\Phi(r)/kT) - 1\}$ is small and negative, representing a positive contribution to $B_2(T)$ (see equation (2.50)) with $B_2 \to 0$ as $T \to \infty$. Conversely, at low temperatures the negative, attractive well dominates the interaction and the integrand $\{\exp(-\Phi(r)/kT) - 1\}$ becomes large and positive. In this case $B_2 \to -\infty$ as $T \to 0$. These features of the classical second virial coefficient are shown in reduced coordinates in Figure 2.7: for classical systems the agreement is seen to be excellent. Evidently there is a temperature, the Boyle temperature, at which $B_2(T)$ is zero, corresponding to collisional sampling in the vicinity of $r = \sigma$ when $\Phi(r)$ is approximately zero.

The Virial Coefficients

For a classical system the second virial coefficient is given by equations (2.36), (2.39) and (2.40) as

$$B_2 = -\frac{1}{2} \int f_{12} \, d2$$

$$= -\frac{1}{2} \int \left\{ \exp\left(-\frac{\Phi(12)}{kT}\right) - 1 \right\} d2 \qquad (2.50)$$

and B_2 is, of course, a function of temperature for realistic pair potentials $\Phi(12)$. As an example of a realistic potential we may take the Lennard–Jones (6–12) pair interaction (Figure 2.6):

$$\Phi(r) = 4\varepsilon \left\{ \left(\frac{\sigma}{r}\right)^{12} - \left(\frac{\sigma}{r}\right)^{6} \right\} \qquad (2.51)$$

where σ is the atomic collision diameter and ε is the well depth. Such a function gives a reasonable representation of the effective pair potential operating in a dense fluid for a number of simple systems, in particular the liquid inert gases.

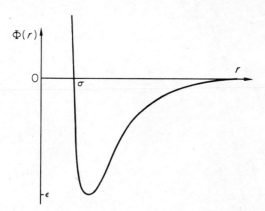

Figure 2.6   The Lennard–Jones (6–12) pair
interaction

It is straightforward to calculate the second virial coefficient for the van der Waals fluid. From equation (2.50)

$$B_2 = 2\pi \int_0^{\sigma} r^2 \, dr - 2\pi \int_{\sigma}^{\infty} \left\{ \exp\left(-\frac{\Phi(r)}{kT}\right) - 1 \right\} r^2 \, dr \qquad (2.52)$$

where the integrand has reduced to -1 over the range $0 < r < \sigma$ (see Figure 2.1). If now we take the high-temperature approximation $\varepsilon/kT \ll 1$, then $\{\exp(-\Phi/kT) - 1\} \sim 1 - \Phi/kT$ and (2.52) becomes

$$B_2 = \frac{2\pi\sigma^3}{3} \left\{ 1 - \frac{3}{s-3} \frac{\varepsilon}{kT} \right\} \qquad (2.53)$$

where we have assumed, as before, $s > 3$ so that the integral converges. The Boyle temperature is therefore

$$T_{\text{Boyle}} = 3\varepsilon/k(s - 3) \qquad (2.54)$$

We may immediately observe that $T_{\text{Boyle}} \to 0$ as $\varepsilon \to 0$ and that there can be no Boyle temperature for a purely repulsive interaction. Moreover, from the magnitude of the index s of the attractive branch of the interaction $(\sigma/r)^s$ we see that $T_{\text{Boyle}} \to 0$ as the attractive interaction strengthens, $s \to \infty$. These two latter observations are of particular relevance in the case of the quantal systems He and H_2 which are seen, from Figure 2.7, to depart significantly from the classical curves. We shall postpone their discussion to a later section, however.

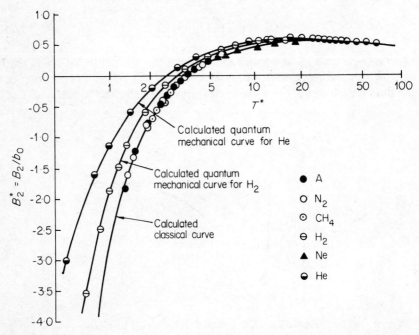

Figure 2.7 The reduced second virial coefficient for classical and quantal systems

In order to calculate the second virial coefficient for a realistic system, the features of $B_2(T)$ will depend parametrically upon the constants σ and ε of the Lennard–Jones potential: the functional dependence in the case of a van der Waals fluid is shown explicitly in equation (2.53). Analysis of the experimental data for the second virial coefficient therefore enables an experimental determination of the force constants σ and ε to be made, and a few of these parameters are listed in Table 2.3.

Exact calculations for the first six virial coefficients for hard spheres and discs have been reported in an earlier section, together with an estimate of $B_7(T)$ and $B_8(T)$. The results for Lennard–Jones systems are, however, less substantial

Table 2.3. Force constants for Lennard–Jones (6–12) systems determined from $B_2(T)$ (after Hirschfelder)

Gas	ε/k (K)	σ (Å)
Ne	34·9	2·78
A	119·8	3·405
Kr	171	3·60
Xe	221	4·10
N_2	95·05	3·698
O_2	118	3·46
CH_4	148·2	3·817
CO_2	189	4·486

and this is to be entirely attributed to the difficulties of numerical evaluation of the cluster integrals for realistic Mayer functions. Nonetheless, the coefficients B_2, B_3, B_4 and B_5 have been determined for the L–J (6–12) potential, and the results are shown in Figure 2.8. It is seen that all the virial coefficients become

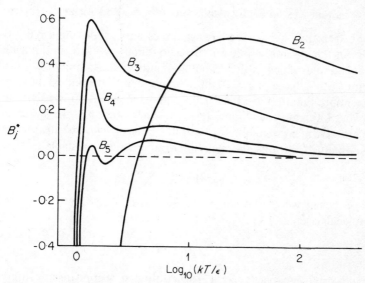

Figure 2.8 The reduced Lennard–Jones virial coefficients B_2 to B_5 calculated on the basis of the cluster integral formulation

large and negative at low temperatures, which raises doubts about the convergence of the virial expansion at low temperatures, even at low densities. Nonetheless, assuming that the virial expansion converges at critical densities, we are able to calculate the critical constants from

$$\left(\frac{\partial P}{\partial V}\right)_{T_c} = 0, \qquad \left(\frac{\partial^2 P}{\partial V^2}\right)_{T_c} = 0 \qquad (2.55)$$

the critical isotherm showing an inflexion at the critical point. Barker has determined these derivatives using a virial expansion terminated at B_3, B_4 and B_5, and observes a convergence upon a critical ratio as more virial coefficients are included (Table 2.4): there is no guarantee, however, that these are the *correct* critical ratios. The critical point is highly singular and the ratios listed in Table 2.4 must be accepted with this reservation.

Table 2.4

Terminating virial coefficient	kT_c/ε	b/V_c	$(PV/kT)_c$
B_3	1·445	0·773	0·333
B_4	1·300	0·561	0·352
B_5	1·291	0·547	0·354

$b = \frac{2}{3}\pi\sigma^3$

Quantum-Mechanical Calculation of the Second Virial Coefficient

We have seen that we cannot neglect the essential indistinguishability of the particles in the assembly and we correspondingly work in the quasi-classical approximation in which explicit account of indistinguishability is taken through the factor $(N!)^{-1}$ in the phase partition function. But to what extent is the quasi-classical approximation justified? That is, to what extent can we simultaneously specify the conjugate coordinates $(\mathbf{p}_i, \mathbf{q}_i)$ of each particle in the system?

A qualitative measure of the applicability of the quasi-classical approximation may be obtained through the Heisenberg uncertainty principle

$$\Delta p \, \Delta q \gtrsim \hbar \qquad (2.56)$$

For a system in which the mean momentum of a gas molecule $\bar{p}$, and mean intermolecular separation $\bar{R}$ satisfies the condition

$$\bar{R}\bar{p} \gg \hbar$$

we should certainly expect a classical description to be appropriate. Equivalently, the mean de Broglie wavelength $\bar{\lambda}$ associated with the particle satisfies

$$\bar{R} \gg \bar{\lambda}$$

classically. If, however, the above inequality does not hold, corresponding to a spatial indeterminacy in the location of the centre of mass of the order of R, then clearly a deterministic classical mechanics is inappropriate. A useful qualitative criterion is the *de Boer parameter* Λ (Table 2.5):

$$\Lambda = \bar{\lambda}/\bar{R} = h/\bar{R}\sqrt{(3mkT)} \qquad (2.57)$$

Table 2.5. De Boer parameter Λ for the noble gases at the triple point

He	0·424
Ne	0·0939
Ar	0·0294
Kr	0·0161
Xe	0·0103

(since $\bar{\lambda} = h/\bar{p}$, $\bar{p}^2/2m = 3kT/2$). Only for systems of small mass, at low temperatures and high densities (small $\bar{R}$) will the departure from the quasi-classical limit be significant.

Numerical Estimate

Consider He^4 at 10 K at atmospheric pressure $\sim 10^6$ dynes/cm^2. From the approximate equation of state the particle number density is

$$\frac{N}{V} = \frac{P}{kT} \sim 2 \times 10^{22} \text{ molecules/cm}^3$$

$$\bar{R} = \left(\frac{V}{N}\right)^{\frac{1}{3}} \sim 8 \times 10^{-8} \text{ cm}$$

$$\bar{\lambda} = h/\sqrt{(3mkT)} \sim 4 \times 10^{-8} \text{ cm}$$

Clearly a classical evaluation of the partition function is inappropriate. A similar calculation for He^4 at 300 K and atmospheric pressure yields

$$\bar{R} \sim 34 \times 10^{-8} \text{ cm}$$

$$\bar{\lambda} \sim 0·6 \times 10^{-8} \text{ cm}$$

In this case, provided the indistinguishability of the particles is acknowledged, the classical approximation ought to be satisfactory. Even for an *ideal* (non-interacting) quantum fluid the statistical class to which the system belongs—Bose–Einstein or Fermi–Dirac—will involve a departure from the classical description. The Pauli repulsion between fermions increases the pressure above the ideal value, whilst the effect of indistinguishability of bosons leads to a decrease in the pressure below that of an ideal classical gas.

We may depict the differences between the Bose–Einstein, Maxwell–Boltzmann and Fermi–Dirac cases by considering a gas of two particles which can be in any of three states, and we shall attempt to enumerate the possible states or *complexions* of the whole gas in the three statistical cases.

Consider first of all the Maxwell–Boltzmann case in which all particles are assumed distinguishable. We therefore discriminate between the two particles

● and ○. The complexions may be enumerated as follows:

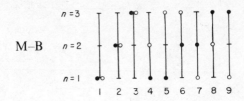

Thus, there are nine complexions in the M–B case for an assembly of two particles in three states, $n = 1, 2, 3$. The assumption of distinguishability enables us to distinguish between configurations 4 and 7, for example, and count these as physically distinct. However, in the *Bose–Einstein* system we no longer assume distinguishability. Clearly this will decrease the number of physically distinct configurations, and we enumerate the complexions as follows:

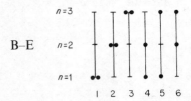

Indeed, there are in the Bose–Einstein case only six physically distinct configurations. If now we apply the further stringent condition of Fermi–Dirac statistics that the wavefunction must be antisymmetric with regard to particle exchange, that is, no two particles can be in the same quantum state, then we find that the number of configurations is further reduced:

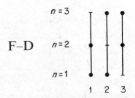

We are now able to make an interesting qualitative observation: Fermi–Dirac systems behave as if there is an additional 'statistical repulsion' between the particles relative to a Maxwell–Boltzmann assembly, whilst a Bose–Einstein system behaves as if there were a 'statistical attraction' developed between the particles. This we can see by determining the dispersion or 'spread' amongst the energy levels in three systems. Thus, for the M–B system the mean of the state separations for the two particles is

$$\text{M–B}: \frac{0+0+0+1+2+1+1+1+2}{9} = \frac{8}{9}$$

whilst for the other two systems

$$\text{B–E}: \frac{0+0+0+1+2+1}{6} = \frac{6}{9}$$

$$\text{F–D}: \frac{1+2+1}{3} = \frac{12}{9}$$

The dispersion amongst the states is therefore 3:4:6 for B–E: M–B: F–D.

The first quantum correction to the ideal-gas pressure results in an increase (decrease) in the case of an ideal Fermi (Bose) gas. Further, the probabilty of finding two particles at a given separation is increased for bosons and decreased for fermions. It may be shown that for a two-particle interaction there is an effective potential (Figure 2.9)

$$\Phi_{\text{exchange}}(r, T) = -kT \ln \left\{ 1 \pm \exp \left(-\frac{2\pi r^2}{\lambda^2} \right) \right\} \qquad (2.58)$$

where the upper sign refers to Bose–Einstein and the lower to Fermi–Dirac statistics. λ is the de Broglie wavelength. Φ_{exchange} arises purely from the symmetry properties of the wavefunction and since it is temperature dependent cannot be regarded as a true interparticle potential. The correction to the

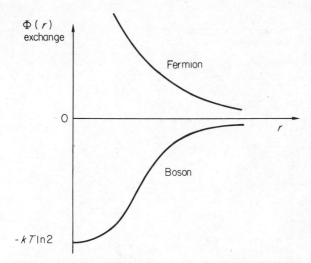

Figure 2.9 The exchange interaction operating between bosons and fermions (equation (2.58)). Although not a true potential, there is an effective statistical attraction between bosons, and repulsion between fermions

second virial coefficient due to exchange effects alone, and therefore applicable to both ideal and non-ideal assemblies, is then

$$B_2(T)_{\text{exchange}} = \pm\frac{1}{2}\left(\frac{\pi\hbar^2}{mkT}\right)^{\frac{3}{2}} \tag{2.59}$$

where the $-(+)$ sign applies to a boson (fermion) system. The Pauli repulsion between fermions increases the pressure above the ideal value, whilst the effect of indistinguishability of bosons leads to a decrease in the pressure below that of an ideal classical gas.

In the quantization of the binary interaction for realistic systems two kinds of solution will arise: a positive *continuum* of unbound levels (other than quantization imposed by boundary conditions) corresponding to high-energy collisional interactions for $kT > |\varepsilon|$, and a degenerate set of discrete levels for $kT < |\varepsilon|$ corresponding to bound states. Both the continuum and discrete states arise as solutions to the radial Schrödinger equation (Figure 2.10). The classical cluster integral defining $B_2(T)$ will therefore be replaced by a sum over the $(2l + 1)$-fold degenerate discrete states plus an integral over the continuum of collisional states. The incident wave suffers a phase shift in the field of the scattering particle: the wave is drawn in, or expelled from the vicinity of the target atom depending whether the negative or positive region of the potential dominates

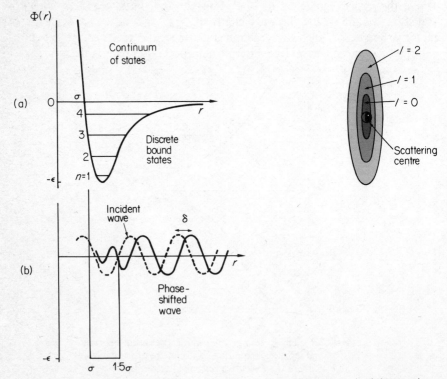

Figure 2.10 (a) The continuum and discrete regions of the quantized interaction. (b) Phase shift of the incident wave in the attractive field of the pair interaction

the interaction, respectively, and this, of course, represents a modification of the dynamics of the collision.

At low temperatures and for short-range potentials the scattering is predominantly s-wave ($l = 0$) and in Figure 2.11 we show the s-wave phase shifts δ_0 for a

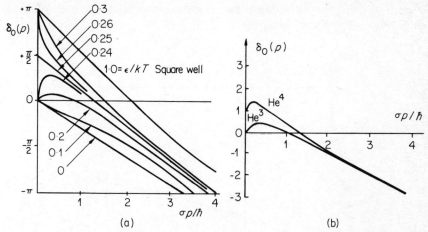

Figure 2.11 (a) Square-well s-wave phase shifts δ_0 as a function of well depth ε/kT. The hard-sphere phase shift is given for $\varepsilon = 0$. (b) The s-wave phase shifts for He^3 and He^4. The statistical repulsion between fermions means that no bound states can develop so that $\delta_0(p) = 0$

square-well interaction, as a function of well depth, together with those for He^3 and He^4. The square-well phase shift varies discontinuously as the well depth is increased. For $\varepsilon = 0$ we have rigid-sphere behaviour. As ε increases discontinuous jumps in the phase shift at $\sigma p/\hbar = 0$ occur, and these correspond to the development of a discrete bound level just at the brim of the well. We see that for low-energy collisions the inward phase shift becomes more and more pronounced with increasing ε as the particle feels the dip in the potential. At high energies the dip is unnoticed and rigid-sphere phase shifts are recovered.

From the curve for He^3 there is a greater Pauli exclusion of the incident wave relative to the He^4 boson phase shift. From the low-energy He^3 phase shift ($\delta_0 = 0$) it may be concluded that no discrete bound states exist, whilst it does appear that there is a discrete state for He^4. At high energies the phase shifts become negative and rigid-sphere-*like*—the Lennard–Jones potential allows for a certain penetration at high energies.

The final expression for the second virial coefficient is

$$B_2(T)_{\text{quantum}} = \pm \frac{1}{2}\left(\frac{\pi\hbar^2}{mkT}\right)^{\frac{3}{2}} - 8\sum_l (2l+1)\left\{\sum_n \exp\left(-\Phi_{nl}/kT\right)\right.$$

$$\underbrace{\qquad}_{\text{exchange}} \qquad\qquad \underbrace{\qquad}_{\text{discrete}}$$

$$\left. + \frac{1}{\pi}\int_0^\infty \frac{d\delta_l}{dp}\exp\left(-p^2/mkT\right)dp\right\} \qquad (2.60)$$

$$\underbrace{\qquad}_{\text{collisional}}$$

$B_2(T)$ for He^4 calculated on this basis is shown in Figure 2.7 and the agreement is seen to be excellent. A comparison of the very-low-temperature calculations for He^3 and He^4 are shown in Figure 2.12. Except for $T \sim 0$K, exchange effects are negligible in comparison with the other terms in equation (2.60). The difference in the $B_2(T)$ curves for He^3 and He^4 may be attributed almost entirely to the absence of discrete contributions in the former case, and the slight difference in the low-energy phase shifts (Figure 2.12).

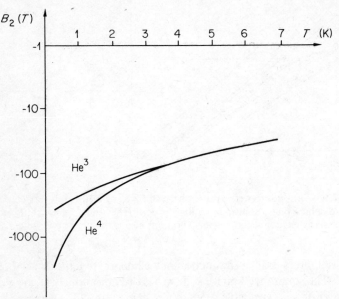

Figure 2.12   The low-temperature second coefficient for He^3 and He^4. Exchange effects become apparent at $T \lesssim 1$K

Completely Ionized Gas

A point charge q will perturb a uniform charge distribution in such a way that beyond a certain radial distance λ^{-1}, the screening or Debye–Hückel length, the field of the central particle is cancelled by the induced redistribution of charge. It is straightforward to show that the Coulombic potential is replaced by a screened Coulombic function $\Phi(r)$ (Figure 2.13)

$$\Phi(r) = \frac{q}{r}\exp\left(-\lambda r\right) \qquad (2.61)$$

where $\lambda^2 = 4\pi\rho q^2/kT$. ρ is the initially uniform density of charges q. The convergence of the cluster integrals arising in the calculation of the virial coefficients depended essentially upon the long-range interaction falling off faster than r^{-3}. For neutral spherically symmetric particles the interaction falls off $\sim r^{-6}$, and difficulties of convergence do not therefore arise. In the case of long-range

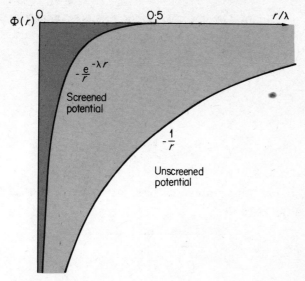

Figure 2.13 Debye–Hückel screening of the Coulomb
potential

electrostatic interactions, such as those arising in a plasma or dilute electrolyte, an alternative approach must be found. For dilute, high-temperature neutral electrostatic assemblies λ is small, and the potential (equation (2.61)) may be expanded:

$$\Phi(r) \sim \frac{q}{r} - q\lambda + \ldots \tag{2.62}$$

The first term represents the Coulomb field of the particle itself, whilst the second is the potential produced by all the other ions in the induced 'cloud' at the point occupied by the ion considered.

The electrical interaction energy of a system of charged particles is $\frac{1}{2} \times$ total number of charges $\times$ potential at the position of the charge due to all the other charges:

$$E = \tfrac{1}{2}(q\rho V)\,\Phi$$

where V is the total volume of the assembly. Thus the *Coulomb* part of the plasma energy is

$$E = -\tfrac{1}{2}q^2\rho\lambda V$$

or, substituting for λ,

$$E = -V(\rho q^2)^{\frac{3}{2}}\sqrt{\frac{\pi}{kT}}$$

The free energy F follows by integrating the thermodynamic relation

$$\frac{E}{kT^2} = -\frac{\partial}{\partial T}\left(\frac{F}{T}\right)$$

giving

$$F = F_{\text{ideal}} - \tfrac{2}{3}V\sqrt{\frac{\pi}{kT}}(\rho q^2)^{\frac{3}{2}}$$

where the constant of integration has been taken as zero so that $F \to F_{\text{ideal}}$ as $T \to \infty$. The pressure is then

$$P = \frac{NkT}{V} - \frac{1}{3V^{\frac{3}{2}}}\sqrt{\frac{\pi}{kT}}(Nq^2)^{\frac{3}{2}} \tag{2.63}$$

where we have replaced ρ by N/V. This expression can be generalized to an assembly of different kinds of ions:

$$P = \frac{kT}{V}\sum_i N_i - \frac{1}{3V^{\frac{2}{3}}}\sqrt{\frac{\pi}{kT}}\left(\sum_i N_i q_i^2\right)^{\frac{3}{2}} \tag{2.64}$$

where the sums run over the various components i of the assembly.

Cell Theories

Difficulties of convergence of the cluster expansion restrict its application to dilute systems, and since the direct evaluation of the N-body configurational partition function $Z_Q(N)$ is impossible some attempt has been made to factorize $Z_Q(N)$ into N independent single-body functions, $Z_Q(1)$. We saw from equation (2.27) that the specific difficulty which arises in the evaluation of the partition function is the representation of the total potential $\Phi_N(1, \ldots, N)$ which depends *simultaneously* on the configuration $(1, \ldots, N)$ of all N particles. If instead we imagine one particle to move in the mean field of its neighbours, then the calculation of its partition becomes straightforward.

At liquid densities there is a certain amount of free volume available to each particle and we may consider the 'wanderer' particle W caged in a potential well by its nearest neighbours, as shown in Figure 2.14. There is some inconsistency in the model in as far as the particles adopt both 'caged' and 'caging' roles— each particle in turn being considered the wanderer. This, however, is an inconsistency we shall have to accept if we are to make any progress.

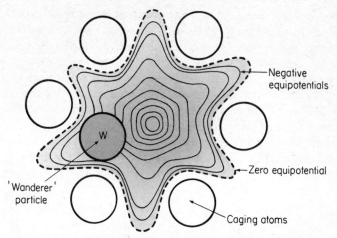

Figure 2.14 Caging of the 'wanderer' particle W in the field of
nearest neighbours

In the *smeared volume model* of Lennard–Jones and Devonshire a certain
simplification is achieved by smearing or 'sphericalizing' the cell potential,
which then becomes purely a function of displacement from the centre of the
cell.

If the mean potential in the assembly is $-\Phi_0$ then the potential of W at some
point r within the cell is $(\Phi(1) - \Phi_0)$ and this will be density dependent. The
single-particle configurational partition function may be written down immediately:

$$Z_Q(1) = \frac{1}{1!} \int \exp\left\{ -\frac{(\Phi(1) - \Phi_0)}{kT} \right\} d\mathbf{1} \qquad (2.65)$$

and for the N-particle assembly, $Z_Q(N) = (Z_Q(1))^N$, i.e. $Z_Q(N) = N \ln Z_Q(1)$.
Then the equation of state is

$$P = kT\left(\frac{\partial \ln Z_N}{\partial V}\right)_T = NkT\left(\frac{\partial \ln Z_Q(1)}{\partial V}\right)_T \qquad (2.66)$$

The problem reduces to the calculation of the cell potential $\Phi(1)$ and the evaluation of $Z_Q(1)$.

Lennard–Jones and Devonshire calculated a series of cell potentials for a
close-packed assembly of particles as a function of density (Figure 2.15). At
high densities a parabolic well characteristic of the solid phase is obtained, the
shape leading to strongly anharmonic behaviour with decreasing density.
Insertion of these cell potentials and evaluation through (2.66) gives the family

of isotherms shown in Figure 2.16. The subcritical isotherms are of the sinuous van der Waals form, whilst those above are monotonic. A comparison of the configurational properties of argon at the triple point ($T = 0.7\varepsilon/k$) with the smeared cell model (Table 2.6) shows the theory to give a better representation of a high-temperature solid than a liquid. This is hardly surprising when we remember that each wanderer is confined permanently within its cage of nearest neighbours—hardly a feature of the liquid state. The discrepancy becomes more acute when we calculate the critical constants: the model of a molecule permanently confined to a cell is quite inappropriate. Various refinements have been made to include the contributions of up to three shells of

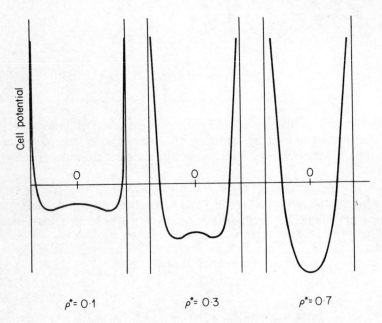

Figure 2.15   Variation of cell potential with density

Table 2.6. Configurational properties of argon at the triple point

	$V/N\sigma^3$	$U/N\varepsilon$	S/Nk	C_v/Nk
Smeared-cell model	1·037	−7·32	−5·51	1·11
Expt. solid argon	1·035	−7·14	−5·33	1·41
Expt. liquid argon	1·186	−5·96	−3·64	0·85

nearest neighbours in the cell potential but a fundamental problem remains: that of the communal entropy.

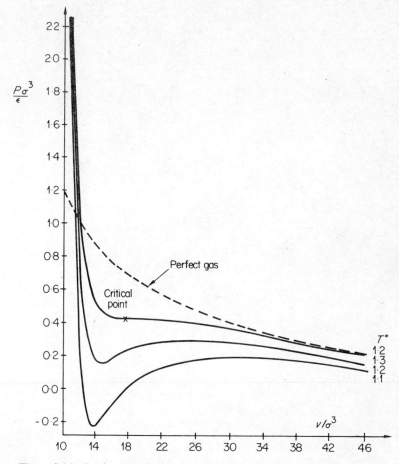

Figure 2.16 Isotherms calculated on the basis of cell theory compared with the perfect-gas curve

The Communal Entropy

The problem of the communal entropy may be most easily considered by comparing the partition function for an assembly of N non-interacting particles in the absence and presence of the cell walls.

(i) If the N particles have access to the entire volume V the total N-body phase-partition function is

$$Z_N = Z_P \left(\frac{V}{N!} \right)^N \tag{2.67}$$

where Z_P is the momentum partition.

46

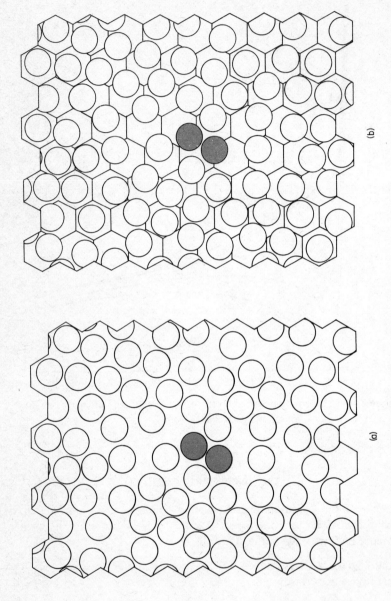

(b)

(a)

Figure 2.17 The 'Wigner–Seitz' single occupancy cells for hard discs with and without the cell walls

(ii) If each particle is restricted to a cell of volume V/N the partition function is

$$_{cell}Z_N = Z_P(Z_Q(1))^N = Z_P(V/N)^N \tag{2.68}$$

Both expressions yield identical expressions for the pressure and internal energy since they involve partial volume and temperature derivatives of $\ln Z_N$. Since the particles are non-interacting the presence of walls is immaterial. The free energy and entropy, depending explicitly upon $\ln Z_N$, are different in the two cases. The quasi-periodic order imposed on the assembly by the cell boundaries can only serve to lower the entropy below the unconstrained value. Thus,

$$\Delta S = S - S_{cell} = k \ln (Z_N/_{cell}Z_N) = k \ln (N^N/N!) = Nk \tag{2.69}$$

i.e. the cell model underestimates the entropy by Nk in the low-density limit. The entropy defect $\Delta S = S - S_{cell}$ is of importance not only in the cell model of liquids but also in the computer simulations. To simulate the effect of an infinite assembly of particles, a series of periodic 'images' surround the central cell in which the evolution of a relatively small number of particles ($\sim 10^3$) is computed (Figure 2.17).

The corresponding discrepancy in the free energy is $\Delta F = -NkT$, and these *communal errors* develop when the particles are constrained and no longer have access to the entire volume V. The problem has provoked considerable discussion. In the crystalline solid partitioning of the free volume does not, of course, lead to a communal entropy defect. In the liquid and gas phases however, the communal problems arise since the particles have unrestricted access to the entire volume. The view that the entire communal entropy makes its appearance at melting, thereby 'explaining' the latent heat of fusion, is by now obsolete (Figure 10.3): the communal entropy is now understood to appear over a density range as shown in Figure 2.18 for hard spheres. The entropy defect ΔS

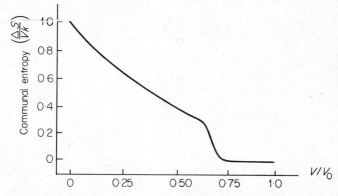

Figure 2.18 The communal entropy defect incurred for a three-dimensional system for a single-occupancy geometry

seems to be an approximately linear function of density up to the phase transition, and suggests that a linear, rather than a constant correction term should be added to simulate the effect of many-body correlation.

Numerous refinements of the cell model have been made to incorporate double and zero cell occupancy together with various concessions to the 'overperiodic' nature of the assembly. Some of these have produced marginal improvements but the model is not sufficiently *a priori* to occupy us further here.

References

C. A. Croxton, *LSP*.

G. H. A. Cole, *An Introduction to the Statistical Theory of Classical Simple Dense Fluids*, Pergamon (1967), Chapters 3, 4.

J. E. Mayer, *Equilibrium Statistical Mechanics*, Pergamon (1968), Chapters 3, 4.

K. Huang, *Statistical Mechanics*, Wiley (1963), Chapters 13, 14.

C. A. Croxton, *Introductory Eigenphysics*, Wiley (1974), Chapter 4.

CHAPTER 3

Equilibrium Structure of Dense Fluids

Introduction

The coefficients in the density expansion of the N-body phase partition function Z_N arise as cluster integrals over groupings of $2, 3, \ldots, N$ particles. The task of numerical evaluation of these cluster integrals rapidly becomes overwhelming, and in consequence only the first five or six cluster integrals have been evaluated. Of course, at low densities the expansion is rapidly convergent and provides an excellent representation of the thermodynamic functions and equation of state of the system. Clearly, there is no possibility at all of evaluating the cluster integrals appropriate to a highly connected liquid, even in the economical Ree–Hoover formalism. Even if the evaluation of the cluster integrals were possible in principle, it may be shown that the radius of convergence of the series restricts its application to densities considerably below that of a liquid. In addition to the thermodynamic properties of the liquid we are, of course, also interested in the *structure* of the system. It is quite apparent that some alternative approach is required.

Instead of attempting to evaluate the total partition function Z_N, the concept of the *correlation function* is introduced. Thus, rather than attempt the total configurational distribution $g_{(N)}(\mathbf{q}^N) = \int \ldots \int f_{(N)}(\mathbf{p}^N, \mathbf{q}^N)\,d\mathbf{p}_1 \ldots d\mathbf{p}_N$ we describe the probability of configurational groupings of two, three, ... particles irrespective of the positions of the remaining $(N-2)$, $(N-3)$, ... particles in the assembly. We shall show that the principal thermodynamic functions can still be calculated, and in addition we now have *structural* information on the liquid.

The *pair distribution function* is of central importance in both the equilibrium and non-equilibrium theory of liquids and it is the relation of this function to the pair potential acting in the liquid which is the main task of the modern equilibrium theory of liquids. This function is directly accessible experimentally, and Zernike and Prins were able to show in 1927 how the pair distribution function is simply related to the Fourier transform of the intensity distribution of scattered radiation. There is the implication here that the structure and thermodynamics of solids and gases could equally well be described in terms of a pair distribution. In principle they could except that other characteristics of these phases suggest alternative approaches and we would not adopt the molecular distribution formalism by choice.

A Formal Relation Between Z_N and the Molecular Distribution

The N-body distribution function $f_{(N)}(\mathbf{p}^N, \mathbf{q}^N)$ is canonically related to the system Hamiltonian $\mathscr{H}(\mathbf{p}^N, \mathbf{q}^N)$ through the Boltzmann expression

$$f_{(N)}(\mathbf{p}^N, \mathbf{q}^N) = \frac{1}{Z_N} \exp\left\{ -\frac{\mathscr{H}(\mathbf{p}^N, \mathbf{q}^N)}{kT} \right\} \tag{3.1}$$

where the phase partition function Z_N enters as a normalizing constant and ensures that the probability of finding the N particles located somewhere in the phase volume is unity. The *configurational projection* of (3.1) may be obtained simply by integrating over the momentum coordinates:

$$\rho_{(N)}(\mathbf{q}^N) = \int f_{(N)}(\mathbf{p}^N, \mathbf{q}^N) \, d\mathbf{p}^N \tag{3.2}$$

$\rho_{(N)}(\mathbf{q}^N)$ represents the spatial number density distribution of the N particles, and for a homogeneous system may be equivalently written $\rho^N g_{(N)}(\mathbf{q}^N)$, where ρ is the uniform number density of the particles and $g_{(N)}$ is the spatial probability distribution. In general we shall be interested in the low-order distribution functions $g_{(h)}(\mathbf{q}^h)$, $h \ll N$, and these are obtained as spatial integral averages over the remaining coordinate variables:

$$\rho^h g_{(h)}(\mathbf{q}^h) = \frac{\rho^N}{(N-h)!} \int \cdots \int g_{(N)}(\mathbf{q}^N) \, d\mathbf{q}_{h+1} \cdots d\mathbf{q}_N \tag{3.3}$$

The factor $1/(N-h)!$ accounts for the physically indistinguishable configurational permutations of the N particles taken h at a time. $g_{(h)}(\mathbf{q}^h)$ is dimensionless and represents the relative probability of finding the subset of h particles in the configuration $\{\mathbf{q}_1, \ldots, \mathbf{q}_h\}$, *relative* that is to the probability of the uniform uncorrelated (random) distribution ($g_{(h)} = 1 \cdot 00$).

If we write down the $(h+1)$th distribution analogous to (3.3) then a recurrence relation is obtained between adjacent orders of correlation function:

$$g_{(h)}(\mathbf{q}_1, \ldots, \mathbf{q}_h) = \frac{\rho}{(N-h)} \int g_{(h+1)}(\mathbf{q}_1, \ldots, \mathbf{q}_{h+1}) \, d\mathbf{q}_{h+1} \tag{3.4}$$

The implication is clear: $g_{(h)}$ may be determined provided we have an explicit knowledge of $g_{(h+1)}$, which in turn requires an explicit knowledge of $g_{(h+2)}$, and so on right up to $g_{(N)}$. This is of course how it should be: all the subset distributions $h < N$ develop in the environment of the total distribution $g_{(N)}$ and if the lower distributions are to be consistent with the total distribution then they are inevitably related. The truncation or termination of the linked hierarchy of equations will be necessary if a closed relation for the distribution functions is to be obtained, but in so doing we are, of course, abandoning consistency with the higher distributions.

From equations (3.1) and (3.2) we may relate the configurational projection to the total potential as follows:

$$\rho^N g_{(N)}(1 \ldots N) = \frac{1}{N! Z_Q} \exp\left\{ -\frac{\Phi_N(1, \ldots, N)}{kT} \right\} \tag{3.5}$$

and from (3.3) the pair distribution, for example, becomes

$$\rho^2 g_{(2)}(12) = \frac{1}{(N-2)! Z_Q} \int \cdots \int \exp\left\{ -\frac{\Phi_N(1, \ldots, N)}{kT} \right\} d3 \ldots dN \tag{3.6}$$

where the configurational partition function is

$$Z_Q = \frac{1}{N!} \int \ldots \int \exp\left\{-\frac{\Phi_N(1,\ldots,N)}{kT}\right\} d\mathbf{1} \ldots d\mathbf{N}$$

(In these purely configurational distributions we take $\{1,\ldots,N\}$ to mean $(\mathbf{r}_1,\ldots,\mathbf{r}_N)$ without ambiguity.) Equation (3.6) represents a formal relation between the pair distribution function $\rho^2 g_{(2)}(12)$ and the total potential. The problem remains as intractible as ever, however, for not only does it contain the unknown total potential $\Phi_N(1,\ldots,N)$, but even contains the configurational partition function, Z_Q.

The Pair Distribution Function

Of the hierarchy of distribution functions, our primary interest concerns $g_{(2)}(12)$, termed the *radial distribution function* (RDF). The RDF of liquid argon is shown in Figure 3.1, and is characteristic of a large number of dense monatomic

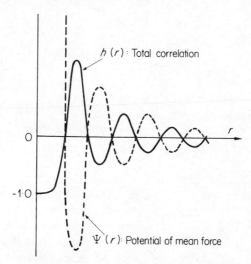

Figure 3.1 The relation between the total correlation function $h(r) = g_{(2)}(r) - 1$ and the potential of mean force $\Psi(r)$

fluids in that $g_{(2)}(12)$ consists of a pronounced first peak located roughly on the pair potential minimum, followed by a number of subsidiary oscillations damping out to unity beyond four or five atomic diameters, this indicating that the subsequent radial structure is uncorrelated to the presence of the origin atom. For spherically symmetric particles the function $g_{(2)}(r)$ gives the relative probability of finding a second particle located a distance r from another particle simultaneously located at the origin. Such distributions are generally taken to apply in the limit $N \to \infty$, $V \to \infty$ whilst $N/V \to \rho$. If we imagine a particle 1 located at the origin of radial coordinates then obviously a second particle is excluded from the region of the short-range repulsive core. If the location of the second particle were independent of the remainder, then the probability maximum would be located on the potential minimum of particle 1. In fact the maximum is displaced slightly inwards due to the anisotropic distribution of molecular collisions, particle 2 shielding particle 1, and vice versa (Figure 3.2).

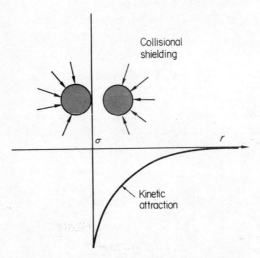

Figure 3.2 Kinetic shielding between two particles resulting in a statistical attraction between hard spheres

A shell of nearest neighbours therefore develops around the origin molecule—the breadth or sharpness of the shell depending on the temperature. From Figure 3.1 we see that there is in fact quite a radial spread about the most probable location. The shell of second-nearest neighbours is then located on the rather imprecisely defined potential minimum of the first shell and so on, the successive probability maxima becoming progressively less well defined until beyond the *correlation length* the structure has become entirely independent of the origin particle and is uniform. (A characteristic of the liquid RDF is that successive peaks become progressively broader.) At lower temperatures the dispersion about the most probable position is correspondingly smaller and the probability maxima are sustained out to greater radial distances; indeed, in the low-temperature crystalline solid phase the periodic structure is sustained over many hundreds of atomic diameters.

This approximate model neglects the weak 'feedback' of molecular correlations into the origin. The structure in fact develops 'self-consistently'. Nevertheless, this represents a simple qualitative description of the development of pair correlation.

We may relate the radially damped pair distribution function to a *hypothetical* radially damped pair potential $\Psi(r)$ through the Boltzmann relation

$$g_{(2)}(r) = \exp \left\{ -\frac{\Psi(r)}{kT} \right\} \qquad (3.7)$$

$\Psi(r)$, the *potential of mean force*, is in fact the radially stacked assembly of pair potentials discussed at the beginning of this section and is shown in Figure 3.1. At small radial separations $\Psi(r) \rightarrow \Phi(r)$ when the pair potential dominates the structure of the fluid, whilst at larger separations the environmental effects of the remaining $(N - 2)$ particles dominate. The potential of mean force may therefore be written as a sum of direct and indirect interactions:

$$\Psi(r) = \Phi(r) + W(r) \qquad (3.8)$$

$\Phi(r)$ represents the direct interaction of particles 1 and 2 whilst the supplement $W(r)$ accounts for the averaged effects of the remainder.

The importance of the effects of the supplementary potential $W(r)$ can be gauged from the hard-sphere RDF shown in Figure 3.3. Since the hard-sphere

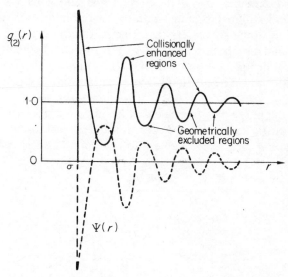

Figure 3.3 Contribution of kinetic shielding to the RDF and the potential of mean force

pair potential has no attractive component the effectively attractive regions of $\Psi(r)$ are to be attributed to the effects of the supplementary interaction $W(r)$. Collisional anisotropy on particle 2 due to interparticle shielding effects simulates an attractive interaction between the particles. This develops to a similar extent in realistic assemblies whilst in a hard-sphere system the maxima can only be attributed to these kinetic effects (Figure 3.2).

The radial distributions shown in Figures 3.1 and 3.3 represent the radial modification of the local uniform structure. Such a distribution is entirely consistent with a uniform particle density in the bulk fluid—from equation (3.4) the single-particle distribution is

$$g_{(1)}(1) = \frac{\rho}{(N-1)} \int g_{(2)}(12)\, d\mathbf{2}$$

or in scalar notation:

$$= \frac{4\pi\rho}{(N-1)} \int g_{(2)}(r)r^2\, dr = 1\cdot00 \tag{3.9}$$

since $4\pi\rho \int g_{(2)}(r)r^2\, dr$ represents the total number of particles in the assembly. In other words, the single-particle density distribution is uniform throughout the bulk system.

Thermodynamic Functions in Dense Fluids

In the case of dilute gases a direct connection to the thermodynamic functions of the system was possible through the cluster expansion of the N-body partition function. We may still make contact with the partition function through equation (3.6), and in doing so express the principal thermodynamic functions in terms of the radial distribution function, $g_{(2)}(12)$. We could equally well express the thermodynamic functions in terms of $g_{(3)}(123)$ or $g_{(4)}(1234)$, but these higher distributions are not experimentally accessible and are difficult to establish analytically.

The role of the phase partition function Z_N in providing a link between the microscopic statistical mechanical description and the macroscopic functions of thermodynamics is clear from Table 3.1. The connection is generally made

Table 3.1

$$F = -kT \ln Z_N$$

$$S = -\left(\frac{\partial F}{\partial T}\right)_V$$

$$P = -\left(\frac{\partial F}{\partial V}\right)_T$$

$$U = F - T\left(\frac{\partial F}{\partial T}\right)$$

through the entropy: the identity of the statistical and thermodynamic entropies is only justified to the extent that there appears to be a macroscopic and experimental coincidence of the thermodynamic entropy with a theoretical microscopic statistical entropy.

The statistical entropy is related to the canonical distribution $f_{(N)}$ as

$$S = -k \langle \ln f_{(N)} \rangle$$

$$= k \ln Z_N + \frac{\langle \mathcal{H} \rangle}{T}$$

where $\langle \mathcal{H} \rangle$ is the phase average of the Hamiltonian. If we designate this phase average as U, the internal energy, and identify $-kT \ln Z_N$ as the free energy F, then the thermodynamic relationship immediately follows

$$F = U - TS$$

This important identification of the free energy forms the basis of nearly all the connections of thermodynamic quantities S, U, P, C_v, etc., with the statistical representation through Z_N.

For a classical, separable total phase partition function we may write

$$\ln Z_N = \frac{3N}{2} \ln \left(\frac{2\pi mkT}{h^2} \right) + \ln Z_Q \tag{3.10}$$

and from the statistical thermodynamic relationship the internal energy

$$U = kT^2 \frac{\partial}{\partial T} [\ln Z_N(V, T)]_V$$

$$= \frac{3}{2} NkT + \frac{1}{N! Z_Q} \int \dots \int \Phi_N \exp \left(-\frac{\Phi_N}{kT} \right) d1 \dots dN$$

$$= \frac{3}{2} NkT + \frac{\rho^2}{2} \iint g_{(2)}(12)\Phi(12) \, d1 \, d2 \tag{3.11}$$

In terms of the radial distribution, from equation (3.6)

$$U = \frac{3}{2} NkT + \frac{\rho^2}{2} \iint g_{(2)}(r)\Phi(r) \, d1 \, d2 \tag{3.12}$$

In writing this we assume pairwise additivity of the total potential. If the molecules exhibit no directional properties so that the system is spherically symmetrical (3.12) reduces to its scalar form:

$$U = \frac{3}{2} NkT + \frac{4\pi \rho^2 V}{2} \int_0^\infty g_{(2)}(r)\Phi(r)r^2 \, dr \tag{3.13}$$

i.e.

$$\boxed{\frac{U}{V} = \frac{3}{2} \rho kT + \frac{4\pi \rho^2}{2} \int_0^\infty g_{(2)}(r)\Phi(r)r^2 \, dr} \tag{3.14}$$

The factor V in the integral (3.13) has appeared from the integration $\int d1$ over all space. The first term in equation (3.14) represents the purely kinetic contribution

56

to the internal energy and in fact is the equipartition value for an equilibrium three-dimensional assembly of N identical particles.

The integral represents the configurational (potential) contribution to the internal energy, and, as we see from Figure 3.4, is essentially negative. The

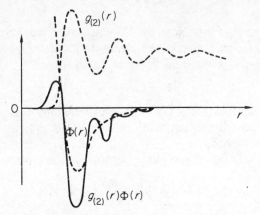

Figure 3.4 The product $g_{(2)}(r)\Phi(r)$

integrand is particularly sensitive to the coincidence or otherwise of the maximum of $g_{(2)}(r)$ and the minimum of $\Phi(r)$. For this reason combination of experimental distribution functions with theoretical pair potentials leads to widely varying estimates of the internal energy. U depends both on density and temperature: explicitly in the case of the kinetic term, and implicitly through the density and temperature dependence of $g_{(2)}(r)$ in the configurational term. The pair potential is assumed temperature- and density-independent; whilst this is generally true, there are certain important exceptions such as the liquid metals where the density and momentum distribution of the conduction electrons decisively affects the nature of the ion–ion interaction.

The Specific Heat, C_v

Assuming the pair potential is temperature independent, the specific heat at constant volume follows from (3.14)

$$C_v = \left(\frac{\partial U}{\partial T} \right)_V$$

$$\frac{C_v}{V} = \frac{3}{2}Nk + \frac{4\pi\rho^2}{2} \int_0^\infty \left(\frac{\partial g_{(2)}(r)}{\partial T} \right)_V \Phi(r) r^2 \, \mathrm{d}r \qquad (3.15)$$

or in the case of a liquid metal

$$\frac{C_v}{V} = \frac{3}{2}Nk + \frac{4\pi\rho^2}{2} \int_0^\infty \left\{ \left(\frac{\partial g_{(2)}(r)}{\partial T} \right)_V \Phi(r) + \left(\frac{\partial \Phi(r)}{\partial T} \right)_V g_{(2)}(r) \right\} r^2 \, \mathrm{d}r \qquad (3.16)$$

Both sodium and argon exhibit a specific heat at constant volume close to the harmonic oscillator value of $3Nk$/mole. Neither system, however, seems to be adequately represented simply as an assembly of harmonic oscillators in the liquid state (Figure 3.5). From the phonon spectra shown in Figure 3.6 liquid

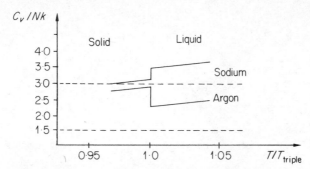

Figure 3.5 Specific heats of sodium and argon in the vicinity of the solid–liquid phase transition. For a three-dimensional assembly of harmonic oscillators $C_v/Nk = 3$

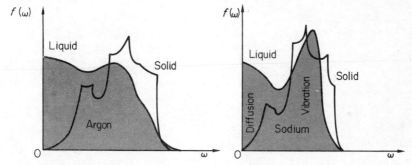

Figure 3.6 Phonon spectra for solid and liquid argon and sodium. Notice the absence of low-frequency diffusive modes in the solid. A qualitative subdivision into low-frequency diffusive and high-frequency vibratory modes is indicated in the case of sodium

sodium appears to retain a more or less well defined vibrational component whilst liquid argon does not. Both systems exhibit a non-zero spectral density at long wavelengths ($\omega \to 0$) and this is characteristic of the liquid phase corresponding to the development of diffusive modes which are obviously absent in the solid phase. A qualitative picture of the motion of a particle as a diffusing oscillator emerges and this is confirmed in the three-dimensional trajectory of an argon atom shown in Figure 3.7.

The diffusive movements can be interpreted as a kind of very-low-frequency vibration: the particle *diffusively* returns to A and therefore appears in the low-frequency region of the spectral density $f(\omega)$ (Figure 3.6). This diffusive return

58

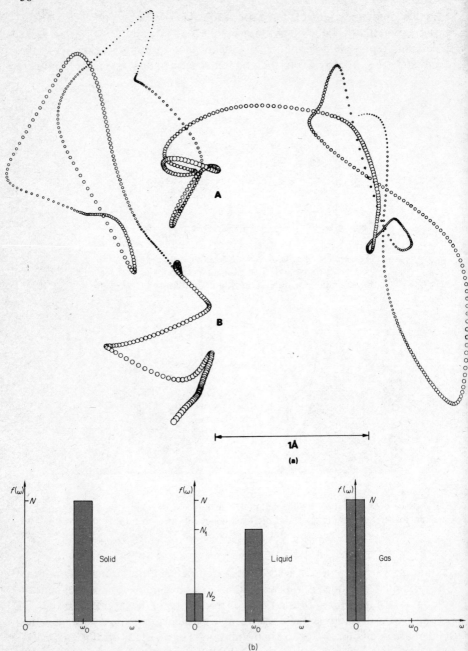

(a)

(b)

Figure 3.7 (a) Molecular dynamics trajectory of a liquid-argon molecule showing vibratory (A) and diffusive (B) components ($\rho = 1.402$ gm cm^{-3}, 178.4 K). (b) Schematic distribution of vibratory (N_1) and diffusive (N_2) modes in solid, liquids and gases, subject to the condition $N = N_1 + N_2$. To account for the general features of the liquid-argon spectral density we require $N_1 : N_2 = 3 : 2$

of the particle occurs in the course of a random walk and *not under any restoring force*. This is important when we come to equipartition of the energy: the high-frequency vibratory modes must be attributed $3k/2 + 3k/2$ per particle accounting for both the kinetic and potential contributions, whilst the purely kinetic low-frequency diffusive modes adopt only $3k/2$ per particle.

It is clear from Figure 3.6 that there is a redistribution of the N normal modes on melting, in particular a substantial replacement of vibratory modes by low-frequency diffusive contributions absent in the crystalline lattice. The distribution of the modes with their associated per-particle equipartition contributions to the specific heat are shown in Table 3.2. If we write a general expression for the specific heat as

$$C_v = \tfrac{3}{2}N_{diff}k + 3N_{vib}k \tag{3.17}$$

substitution of the values in Table 3.2 yields the various specific heats. In particular, the specific heat of the liquid will decrease on melting if $N_1 < N$

Table 3.2

	$N_{diff}(\tfrac{3}{2}kT)$	$N_{vib}(3kT)$
Solid	0	N
Liquid	N_2	N_1
Gas	N	0

and in the case of argon this is seen to be the case. In the case of sodium the discussion is less simple due to the discontinuous variation of the ionic interaction on melting (equation (3.16)).

It has been assumed throughout the discussion that the normal mode representation of the assembly is adequate. In fact the motions are strongly anharmonic which implies pronounced phonon coupling in the assembly—a feature neglected in the description above.

Theoretical calculations for the liquid inert gases on the basis of equation (3.15) are not yet reliable. There is little experimental data available for $(\partial g_{(2)}(r)/\partial T)_V$, and any theoretical attempt to specify this derivative formally involves the higher-order distribution $g_{(3)}$. The temperature dependence of any of the approximate theories of the pair distribution function can in principle be determined, but the results are unlikely to be useful at this stage.

The Equation of State

The equation of state, or pressure, is related to the volume derivative of the configurational partition function as follows

$$P = kT\frac{\partial}{\partial V}(\ln Z_Q(V, T))_T \tag{3.18}$$

The total potential $\Phi_N(1, \ldots, N)$ is, of course, a function of the volume, and must appear explicitly in (3.18). In the case of dilute gases we were able to effect a density expansion for Z_Q in which the explicit volume dependence appeared termwise.

If we make a change in the radial variable, thus

$$r = r'V^{\frac{1}{3}} \tag{3.19}$$

then we may write r in terms of V.

$$Z_Q = \frac{V^N}{N!} \int \ldots \int \exp\left(-\frac{\Phi_N(1, \ldots, N)}{kT}\right) d1 \ldots dN$$

$$\ln Z_Q = \ln V^N + \ln\left\{\frac{1}{N!} \int \ldots \int \exp\left(-\frac{\Phi_N}{kT}\right) d1 \ldots dN\right\}$$

From the first term in $\ln Z_N$, equation (3.18) yields the kinetic contribution to the pressure, NkT/V. The second term involves the differential $\partial \Phi_N / \partial V$, and assuming pairwise additivity we have

$$\frac{\partial \Phi_N}{\partial V} = \sum_{i<j} \sum \frac{\partial \Phi(r_{ij})}{\partial r_{ij}} \cdot \frac{r_{ij}}{3V}$$

The transformation (3.19) is now reversed, and by an analogous procedure to that of the calculation of U, we obtain

$$P = \frac{NkT}{V} - \frac{1}{Z_Q} \int \ldots \int \exp\left(-\frac{\Phi_N(1, \ldots, N)}{kT}\right) \frac{\partial \Phi_N}{\partial V} d1 \ldots dN$$

$$= \frac{NkT}{V} - \frac{\rho}{6} \iint g_{(2)}(r_{12})r_{12} \frac{d\Phi(r_{12})}{dr_{12}} d1\, d2$$

Since the integral is independent of the location of the pair, and for spherical particles

$$P = \rho kT - \frac{4\pi\rho^2}{6} \int_0^\infty g_{(2)}(r)\frac{d\Phi(r)}{dr} r^3\, dr \tag{3.20}$$

In the limit of low densities this expression reduces to its kinetic component as it does for non-interacting systems. The configurational component gives the contribution arising in condensed assemblies and, subject to the restrictions of the pair approximation, is applicable over a wide range of conditions.

The pressure equation of state provides a very severe test of both the pair distribution and the pair potential since $(d\Phi(r)/dr)$ is a very rapidly varying function in the vicinity of the first peak of $g_{(2)}(r)$. Small errors in the location of the peak or in the assumed form for the pair potential can lead to serious errors in the estimate of the pressure which, in equation (3.20), is determined as the difference between two large quantities.

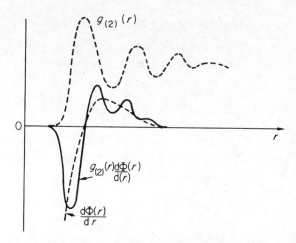

Figure 3.8 The product $g_{(2)}(r)\,d\Phi(r)/dr$

An equivalent equation, whose derivation will not be given here, relates the isothermal compressibility $kT(\partial\rho/\partial P)_T$ to the *total correlation* $h(r) = g_{(2)}(r) - 1$:

$$\frac{1}{kT}\left(\frac{\partial P}{\partial\rho}\right)_T = 1 + 4\pi\rho \int_0^\infty h(r)r^2\,dr \qquad (3.21)$$

For a structureless system $g_{(2)}(r) = 1\cdot00$ and the right-hand side of (3.20) reduces to zero and we regain the ideal equation of state. Another particularly important form of the *compressibility equation of state* arises in combination with the Ornstein–Zernike equation (3.48):

$$\frac{1}{kT}\left(\frac{\partial P}{\partial\rho}\right)_T = 1 - 4\pi\rho \int_0^\infty c(r)r^2\,dr \qquad (3.22)$$

where $c(r)$ is the *direct correlation*. Equations (3.20) and (3.22) are, of course, reformulations of each other and should yield identical isotherms. We shall find, however, that except at the lowest fluid densities there is a significant discrepancy between the pressure and compressibility equations of state which may be directly attributed to approximations made in accounting for many-body effects. In other words, the approximations engender inconsistencies amongst the correlations $c(r)$ and $h(r)$ and this is reflected in the inconsistency in the equations of state. Attempts have been made to 'force' self-consistency on the equations: these 'self-consistent approximations' will be discussed in some detail later.

From Table 3.1 it is seen that the free energy and entropy depend upon the total phase partition function Z_N, rather than on its partial derivatives. There is no means at present of evaluating the partition function directly and we cannot express F and S simply in terms of the pair distribution function. It is possible to form density expansions of these quantities, but difficulties of convergence make this approach inappropriate at liquid densities.

Equations for the Pair Distribution Function

There is, as we have seen, no difficulty in establishing a formal relation between the pair distribution and the pair potential (equation (3.6)):

$$\rho^2 g_{(2)}(12) = \frac{1}{(N-2)! Z_Q} \int \ldots \int \exp \left\{ -\sum_{i>j} \sum \Phi(ij)/kT \right\} d3 \ldots dN \quad (3.23)$$

Another feature of the distribution functions is their membership of a linked hierarchy of distribution equations (equation (3.4)), and for the pair distribution in particular:

$$g_{(2)}(12) = \frac{\rho}{(N-2)} \int g_{(3)}(123) \, d3 \quad (3.24)$$

The first class of integrodifferential equations relating the pair distribution to the pair potential eliminate the Z_Q-dependence in equation (3.23) by considering the spatial variation of the distribution $\nabla g_{(2)}(12)$, rather than the distribution itself. The coupled hierarchical relationship (3.24) remains, however, and some means of terminating the dependence of $g_{(2)}(12)$ on higher-order distributions has to be found. The 'Kirkwood closure' does terminate the hierarchy and provides a closed expression (i.e. in terms of $g_{(2)}(12)$ only) for the pair distribution, but destroys consistency with the higher-order distributions in doing so. The Kirkwood closure is adopted for reasons of mathematical convenience rather than physical significance and to this extent is physically unsatisfactory.

A second and more recent class of equations attempts to resolve the total pair correlation between particles into a direct and indirect effect in much the same way that the potential of mean force $\Psi(12)$ is resolved into its direct and indirect components (equation (3.8): now we work at the level of correlation rather than molecular interaction.

$$\Psi(12) = \Phi(12) + W(12)$$
$$\text{total} = \text{direct} + \text{indirect} \quad (3.25)$$

Thus, the *direct correlation* between particles 1 and 2 is imagined to be supplemented by an *indirect correlation* transmitted from 1 through various chains of direct correlation to particle 2. Provided a simple analytic expression for the direct correlation can be found, the total correlation can be expressed as a series of cluster diagrams, rather like those of the virial expansion. In this case there is extensive cancellation amongst the diagrams and approximate diagrammatic sums may be obtained for the direct correlation, and ultimately the pair distribution determined.

Both these routes, which are formally very similar, lead to theoretical pair distributions on the basis of which the various thermodynamic functions may be determined (equations (3.14), (3.15), (3.20), etc.). A particularly severe test of the distribution function is the calculation of the equation of state which depends very sensitively on the features of the principal peak in $g_{(2)}(r)$. We shall therefore generally make comparisons in terms of the equation of state for dense fluid systems.

The Born–Green–Yvon Equation (BGY)

A simple physical derivation of the Born–Green–Yvon (BGY) equation may be given as follows. Consider a triplet of particles at the points **1**, **2** and **3**: the pair distribution may be expressed in terms of the hypothetical *potential of mean force*, $\Psi(12)$ (equation (3.7)):

$$g_{(2)}(12) = \exp\left\{-\frac{\Psi(12)}{kT}\right\} \tag{3.26}$$

Now, the net force on particle 1 will be the *direct* effect of particle 2, plus the net effect of all the other particles. We may consider particle 3 to be representative of the rest of the fluid and average its effect on particle 1. This will involve particle 3 ranging over the available volume, subject, of course, to its remaining correlated with 1 and 2. Particle 3, in other words, cannot range over *all* space without neglecting the presence of 1 and 2. Indeed, if it did so the distribution of forces on 1 due to the rest of the fluid would be spherically symmetric and particles 1 and 2 would interact as if they were in a dilute gas.

The net force on particle 1 is, therefore

$$\rho g_{(2)}(12)\nabla\Psi(12) = -\rho g_{(2)}(12)\nabla\Phi(12) - \rho^2 \int \nabla\Phi(13)g_{(3)}(123)\,d3 \tag{3.27}$$

The triplet distribution $g_{(3)}(123)$ has arisen since particle 3 is subject to correlation with particles 1 and 2.

Combining equations (3.26) and (3.27) we have the BGY equation:

$$\boxed{-kT\nabla g_{(2)}(12) = g_{(2)}(12)\nabla\Phi(12) + \rho \int \nabla\Phi(13)g_{(3)}(123)\,d3} \tag{3.28}$$

and generally,

$$-kT\nabla g_{(h)}(\mathbf{r}^h) = g_{(h)}(\mathbf{r}^h) \sum_{i=2}^{N} \nabla\Phi(1, i) + \rho \int \nabla\Phi(1, h+1)g_{(h+1)}(\mathbf{r}^{h+1})\,d(h+1)$$

$$\tag{3.29}$$

Both these equations are exact to the extent of the assumed pairwise additivity of the total potential. Both equations show the expected hierarchical dependence of $g_{(h)}(\mathbf{r}^h)$ upon $g_{(h+1)}(\mathbf{r}^{h+1})$. Consequently (3.28) and (3.29) do not represent closed expressions which may be solved for the pair distribution: thus, as it stands, (3.28) represents little more than a formal relationship between $g_{(2)}(12)$

and $g_{(3)}(123)$ and brings us no nearer a tractable expression for the pair distribution. An analogous expression relating $g_{(3)}$ and $g_{(4)}$ may be set up, and indeed this represents the approach of Cole and Fisher. A determination of $g_{(3)}$ in terms of $g_{(4)}$ merely postpones rather than overcomes the problem of breaking the linked chain of equations however, and not until we terminate the hierarchy and effect some form of 'closure' on the equations can we hope to have anything other than a formal solution to our problem.

The usual method of termination is to express $g_{(h+1)}(\mathbf{r}^{h+1})$ in terms of $g_{(h)}(\mathbf{r}^{h})$, or for present purposes, $g_{(3)}(123)$ in terms of $g_{(2)}(12)$. Of course, in adopting this approximate representation we forfeit consistency with the higher-order distributions. In other words, $g_{(2)}(12)$ determined with such a closure device is inconsistent with the exact $g_{(3)}(123)$.

The Kirkwood Superposition Approximation

Kirkwood suggested the first closure for $g_{(3)}(123)$ in terms of $g_{(2)}$: this 'superposition approximation' enables the BGY equation to be written entirely in terms of the pair distribution:

$$\boxed{g_{(3)}(123) \sim g_{(2)}(12)g_{(2)}(23)g_{(2)}(31)} \qquad (3.30)$$

Physically this amounts to saying that in the triplet of particles, 1 and 2 correlate independently of the presence of 3, similarly 2 and 3 independently of 1, and 3 and 1 independently of 2. Only in certain limiting cases can this be a reasonable approximation. At large separations appropriate to low densities the particles are effectively uncorrelated and as $r_{ij} \to \infty$ so $g_{(2)}(ij) \to 1.00$. In this case the exact and the superposition approximation yield the same limiting value (Figure 3.9a):

$$g_{(3)}(123) \to g_{(2)}(12) \to 1.00 \quad \text{as } r_{ij} \to \infty$$

When one atom is distant from the other two (Figure 3.9b)

$$g_{(3)}(123) \to g_{(2)}(12) \quad \text{as } g_{(2)}(13), \quad g_{(2)}(23) \to 1.00$$

and this represents the second case of limiting validity, neither of which appears appropriate in the high-density strongly coupled liquid.

The essential inconsistency between the two- and three-body distributions in the superposition approximation can be seen from the recurrence relation

$$g_{(2)}(12) = \frac{\rho}{N-2} \int g_{(3)}(123) \, \mathrm{d}3 \qquad (3.31)$$

which, in the superposition approximation, may be written

$$\frac{N-2}{\rho} = \int g_{(2)}(23)g_{(2)}(31) \, \mathrm{d}3$$

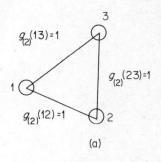

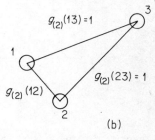

Figure 3.9 Limiting configurations for an exact superposition approximation

which is obviously incorrect since the right-hand side of (3.31) is a function
of the separation 12, whilst the left-hand side is a constant.

The superposition approximation may, nevertheless, be shown to be a simple
mathematical expression representing a cluster sum containing contributions
to all orders of density. Its insertion in (3.28) yields a closed non-linear integrodifferential equation for the pair distribution

$$-kT\nabla \ln g_{(2)}(12) = \nabla\Phi(12) + \rho \int \nabla\Phi(13)g_{(2)}(23)g_{(2)}(31)\,d3 \qquad (3.32)$$

which is solved subject to the boundary condition

$$g_{(2)}(12) \to 1 \quad \text{as } r_{12} \to \infty$$

Equations such as (3.32) are generally solved numerically by an iterative
technique. An approximate form for the distribution function (but nevertheless
having the correct limiting forms $g_{(2)}(0) = 0$, $g_{(2)}(\infty) = 1\cdot0$) is substituted
behind the integral. The integral is then evaluated and the new expression for
$g_{(2)}(12)$ on the left-hand side forms the basis for the next iteration. This process
is repeated until $g_{(2)}(12)$ converges on some final form which does not change
with further iteration.

A comparison of the exact and BGY radial distributions for a high-density
Lennard–Jones fluid is made in Figure 3.10. The agreement is bad and this
must be attributed to the superposition approximation. In particular the
principal peak is seriously in error. Not surprisingly, judged by its prediction
of thermodynamic quantities, the BGY theory is the least successful of all the

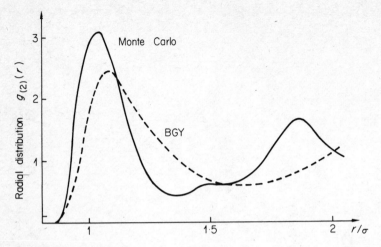

Figure 3.10 Comparison of the simulated and BGY radial distributions
for liquid argon

theories. In one respect, however, it is unique: for hard spheres Kirkwood,
Maun and Alder demonstrated that (3.32) ceases to yield convergent or physically acceptable solutions beyond some limiting density. Molecular dynamic
computer simulations of hard-sphere systems yielded quite clear evidence of
a solid–fluid phase transition at about the density the BGY equation ceased
to be integrable, and Levesque has found similar instabilities in solving the
BGY equation for more realistic potentials.

Whether this happens to be an accidental feature of the equation, or of the
superposition approximation, or whether it does signal the phase transition,
has been debated ever since with no rigorous answer.

Adequacy of the Superposition Approximation

The superposition approximation having limiting validity at large separations
only would appear inappropriate to dense fluid assemblies. In fact the approximation is surprisingly successful, and this, coupled with the suggestion of a
BGY phase transition has stimulated considerable interest in refining the
approximation.

A *direct* experimental measurement of $g_{(3)}(123)$ is not possible, although there
is some indirect evidence. $g_{(3)}(123)$ may, however, be observed directly in
computer simulation and the equilateral configuration $g_{(3)}(x, x, x)$ measured
directly. If the Kirkwood superposition approximation were exact the ratio

$$[g_{(3)}(x, x, x)]^{\frac{1}{3}}/g_{(2)}(x) \tag{3.33}$$

would be everywhere unity. The results for dense hard-sphere and Lennard–
Jones fluids are shown in Figure 3.11. For hard spheres the ratio $[g_{(3)}(x, x, x)]^{\frac{1}{3}}/$
$g_{(2)}(x)$ is unity to within ~ 10 per cent, and whilst the scatter about unity is

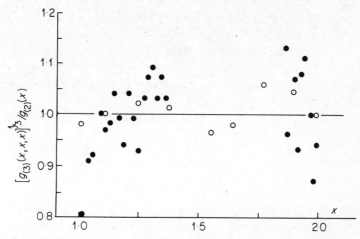

Figure 3.11 Test of the adequacy of the superposition approximation. If exact, the ratio $[g_{(3)}(x, x, x)]^{\frac{1}{3}}/g_{(2)}(x)$ would be unity, the scatter of points for hard spheres (○) and for liquid argon (●) is seen to be surprisingly satisfactory

somewhat greater for Lennard–Jones fluids, the approximation is nevertheless surprisingly accurate. Indirect experimental evidence confirms this conclusion in the case of argon, although it does appear that the discrepancy remains even at large separations. An experimental determination for rubidium seems to suggest a long-range oscillatory supplement to the superposition product is necessary for an adequate description of $g_{(3)}(123)$.

A Modification of the Superposition Approximation

The qualitative correctness of the BGY equation, in as far as it ceases to be integrable beyond some limiting density, suggests that we might profitably investigate possible improvement of the Kirkwood superposition approximation. Bearing in mind the surprisingly satisfactory nature of the pair approximation to $g_{(3)}(123)$ at liquid densities, it is to be anticipated that improvement of the superposition approximation will lead to a significant improvement of the radial distribution function and the equation of state.

Meeron and Salpeter have independently shown that the exact expression for the triplet distribution is

$$g_{(3)}(123) = g_{(2)}(12)g_{(2)}(23)g_{(2)}(31) \exp \left\{ \sum_{n=1}^{\infty} \rho^n \delta_{n+3}(123) \right\} \qquad (3.34)$$

where the coefficients $\delta_{n+3}(123)$, termed 'simple 123 irreducible clusters' by Salpeter, give the contributions to correlations between the particles 1, 2 and 3 and n field points.

In terms of the potential of mean force, (3.34) amounts to

$$\Psi(123) = \Phi(12) + \Phi(23) + \Phi(31) - kT\left\{\sum_{n=1}^{\infty} \rho^n \delta_{n+3}(123)\right\} \qquad (3.35)$$

where the final term represents the supplement to the three-body potential estimated in the pair approximation.

The coefficient δ_4 represents the interaction of the triplet with *one* field point:

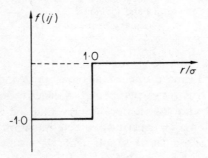

$$\delta_4(123) = \quad\equiv \int f_{14} f_{24} f_{34}\, \mathbf{d4} \qquad (3.36)$$

where f_{ij} is the usual Mayer f-function. In the case of hard spheres f_{ij} is the negative step function shown in Figure 3.12. Since the f-functions are either -1 or 0, $\delta_4(123)$ is evidently negative and its form is shown in Figure 3.13.

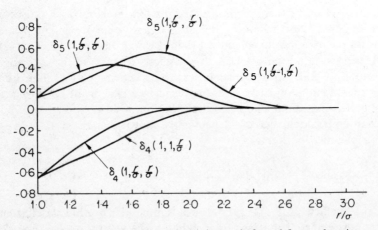

Figure 3.12 The hard-sphere Mayer f-function

Figure 3.13 The hard-sphere cluster integrals δ_4 and δ_5 as a function of configuration

$\delta_5(123)$ represents the interaction of the triplet with *two* field points. The number of diagrams involved in its definition is substantially increased, and these are shown in Figure 3.14.

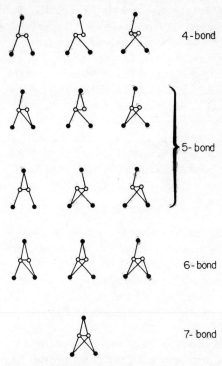

Figure 3.14 Components of the coefficient δ_5

The diagrams involved in $\delta_5(123)$ are grouped in Figure 3.14 according to the number of f-bonds in the cluster, and for hard spheres the diagrammatic contributions are positive or negative according as the number of bonds is even or odd. Clearly there is extensive cancellation amongst the thirteen diagrams and the eventual contribution of $\delta_5(123)$ is weak.

The numerical evaluation of the higher coefficients is prohibitively expensive in terms of computer time and the simplest Padé approximant to the series is formed (see equation (2.47))

$$\sum_{n=1}^{\infty} \rho^n \delta_{n+3}(123) \sim \frac{\rho \delta_4(123)}{1 - \rho \delta_5(123)/\delta_4(123)} \qquad (3.37)$$

and this would be expected to give a better representation to the Salpeter series that the 'straight' linear combination $\rho \delta_4 + \rho^2 \delta_5$. The cancellation amongst the higher terms in the Salpeter series is simulated in the Padé approximant by a 'scaling down' of the first term $\rho \delta_4(123)$, whilst the linear combination represents the cancellation rather less completely. The results are discussed in Chapter 4: the Padé approximant to the distribution $g_{(3)}(123)$ represents a considerable improvement on BGY in the superposition approximation.

This approach has been extended to include Lennard–Jones fluids, and the results are in substantial agreement with the machine simulations.

We can now understand the surprisingly satisfactory nature of the Kirkwood superposition approximation. The extensive cancellation amongst higher-order terms results in a correction to the superposition approximation which is considerably less dependent upon density than we might have first thought. The cancellation arises from the nature of the hard-sphere f-bond which causes the coefficients $\delta_{n+3}(123)$ to alternate in sign. The cancellation is, however, rather more complete in a rigid-sphere fluid than in a Lennard–Jones system where the f-bonds and hence $\delta_{n+3}(123)$ are not exclusively positive or negative over their entire range. This accounts for the slightly less satisfactory nature of the superposition product in the LJ case (Figure 3.11).

Closure on $g_{(4)}(1234)$

Instead of forming the closure on the hierarchy of distributions at third order (i.e. on the triplet of particles), presumably some improvement in $g_{(2)}(12)$ should result if the closure is postponed until *fourth* order. Thus we attempt to apply closure in the subspace of molecular quadruplets rather than triplets and this will inevitably involve the distributions $g_{(3)}(123)$ and $g_{(4)}(1234)$: this provides a natural extension of the BGY equation in the spirit of the superposition approximation.

The equations which result from the various prescriptions for closure at fourth order are incomparably more difficult to solve than the ordinary BGY equation. In fact no numerical solutions have been obtained at liquid densities, although Cole has been able to estimate the first-order correction to the Kirkwood superposition approximation in a density expansion for dilute gases. We shall only briefly consider the various closures which have been proposed: for a more detailed treatment the reader is referred to LSP.

The closure of the quadruplet distribution $g_{(4)}(1234)$ may be effected in several ways. Cole, for example, sets

$$g_{(4)}(1234) \equiv g_{(3)}(123)g_{(2)}(14)g_{(2)}(24)g_{(2)}(34) \qquad (3.38)$$

and takes the Kirkwood superposition approximation for $g_{(3)}(123)$. This form is asymmetric in as far as it has the effect of singling out particle 4. No solutions at liquid densities have been obtained, but the first-order correction to the three-body superposition approximation for dilute gases has low-density Salpeter form:

$$g_{(3)}(123) = g_{(2)}(12)g_{(2)}(23)g_{(2)}(31)[1 + \rho\alpha k_{(3)}(123) + \ldots] \qquad (3.39)$$

where $k_{(3)}(123)$ is a known simple function of the set of three particles and α is a numerical factor determined by this procedure.

Fisher has proposed closure at fourth order, but accounts exactly for the triplet distributions. Thus,

$$g_{(3)}(123) = g_{(2)}(12)g_{(2)}(23)g_{(2)}(31)T_{(3)}(123) \qquad (3.40)$$

and

$$g_{(4)}(1234) = g_{(3)}(123)g_{(3)}(234)g_{(3)}(341)g_{(3)}(142) \qquad (3.41)$$

$T_{(3)}(123)$ is the *indirect* triplet correlation representing the interaction of the triplet 123 with a 4th, 5th, ... particle and is entirely analogous to the Salpeter series (3.34). Equation (3.41) has the advantage of being symmetric with respect to the constituent particles. We are now able to write down two coupled BGY equations, one relating $g_{(2)}$ to $g_{(3)}$ and the other $g_{(3)}$ to $g_{(4)}$. Using the closures (3.40) and (3.41) we finally obtain two coupled equations entirely in terms of $g_{(2)}$ and $T_{(3)}$. These two equations are then solved 'simultaneously' subject to the boundary conditions

$$\left. \begin{array}{c} g_{(2)}(12) \to 1 \\ T_{(3)}(123) \to 1 \end{array} \right\} \begin{array}{l} \text{as any one of the} \\ \text{particles} \to \infty \end{array}$$

which ensure that the distributions are correctly normalized. Again, no solutions are available at liquid densities.

Abe's Series Expansion of the BGY Equation

A series expansion of the BGY equation in the superposition approximation which is in a particularly convenient form for numerical solution is that of Abe. The series develops as a successive resubstitution process as follows.

In the BGY equation

$$-kT\nabla g_{(2)}(12) = g_{(2)}(12)\nabla\Phi(12) + \rho \int g_{(2)}(12)g_{(2)}(23)g_{(2)}(13)\nabla\Phi(13)\,d3 \qquad (3.42)$$

we may substitute for $g_{(2)}(13)\nabla\Phi(13)$ from

$$-kT\nabla g_{(2)}(13) = g_{(2)}(13)\nabla\Phi(13) + \rho \int g_{(2)}(13)g_{(2)}(34)g_{(2)}(14)\,d4$$

whereupon (3.42) becomes

$$-kT\nabla g_{(2)}(12) = g_{(2)}(12)\nabla\Phi(12) + \rho \int g_{(2)}(12)g_{(2)}(23)$$

$$\times \left[-\nabla g_{(2)}(13)kT - \rho \int g_{(2)}(13)g_{(2)}(34)g_{(2)}(14)\nabla\Phi(14)\,d4 \right] d3$$

Repeating this process for $g_{(2)}(14)\nabla\Phi(14)$:

$$-kT\nabla g_{(2)}(12) = g_{(2)}(12)\nabla\Phi(12) - kT\rho \int g_{(2)}(12)g_{(2)}(23)\nabla g_{(2)}(13)\, d3$$

$$+ kT\rho^2 \iint g_{(2)}(12)g_{(2)}(23)g_{(2)}(13)g_{(2)}(34)$$

$$\times \left[-\nabla g_{(2)}(14)kT - \rho \int g_{(2)}(14)g_{(2)}(45)g_{(2)}(15)\nabla\Phi(15)\, d5 \right]$$

$$\times d4\, d3 \qquad (3.43)$$

The general expression of this series is

$$\boxed{\begin{aligned}
-kT\nabla \ln g_{(2)}(12) &- \nabla\Phi(12) = kT \sum_{n=1}^{\infty} (-1)^n \rho^n \\
&\times \int_3 \ldots \int_{n+2} \{\nabla g_{(2)}(1, n+2)\} g_{(2)}(n+1, n+2) \\
&\times \prod_{i=3}^{n+1} \{g_{(2)}(1, i)g_{(2)}(i-1, i)\}\, d3 \ldots d(n+2)
\end{aligned}} \qquad (3.44)$$

The simplicity of (3.44) may be more immediately realized by representing the Abe expansion diagrammatically

$$-kT\nabla \ln g_{(2)}(12) - \nabla\Phi(12)$$

$$= kT \left[-\rho \, \text{⟨diagram⟩} + \rho^2 \, \text{⟨diagram⟩} - \rho^3 \, \text{⟨diagram⟩} + \rho^4 \, \text{⟨diagram⟩} - \ldots \right] \qquad (3.45)$$

where $\sim$ represents the bond $\nabla g_{(2)}(1, n+2)$ and $-$ represents the $g_{(2)}$ bonds. The field points ○ represent the integration variables and range over all space whilst the root points ● are held fixed at the separation r_{12}. The series (3.45) is computationally convenient since numerical evaluation of the kth term greatly simplifies evaluation of the $(k+1)$th. Thus a knowledge of ⟨diagram⟩ for all root point separations means that

$$\text{⟨diagram⟩} \rightarrow \text{⟨diagram⟩} \quad \text{where} \quad \text{⟨diagram⟩} \equiv \text{⟨diagram⟩}$$

similarly,

$$\text{⟨diagram⟩} \rightarrow \text{⟨diagram⟩} \quad \text{where} \quad \text{⟨diagram⟩} \equiv \text{⟨diagram⟩}$$

and so on.

The cluster diagrams arising in the BGY series expansion (3.45) can be represented by a simple closed loop and a complete set of cross-links radiating from particle 1. Abe has obtained (3.45) in closed form by neglecting the cross-links:

this amounts to setting $g_{(2)}(1, i) = 1.00$, i.e. particles 1 and i are uncorrelated. The resulting solution is

$$-\frac{\Phi(r)}{kT} = \ln g_{(2)}(r) - \frac{1}{(2\pi)^3 \rho} \int \frac{[1 - S(k)]^2}{S(k)} e^{i\mathbf{k}\cdot\mathbf{r}} \, d\mathbf{k} \qquad (3.46)$$

where $S(k)$ is the *structure factor* and is simply related to the Fourier transform of $[g_{(2)}(r) - 1]$. Equation (3.46), interestingly enough, may be shown to be identical to the HNC approximation; the connection will be established in a later section.

Direct and Indirect Correlation

A second and more recent route to the determination of fluid structure is through the concepts of direct and indirect correlation. We have seen how the potential of mean force $\Psi(12)$ may be resolved into a direct component $\Phi(12)$ which is the straightforward pair potential and a supplement $W(12)$ which accounts for the modifying effect of the remaining $(N - 2)$ particles.

We now make an analogous resolution in terms of the *correlation* rather than the potential, although the resolution of the total correlation into a direct and indirect component will, of course, have consequences for the potential but these will in general be different to the breakdown in equation (3.25).

The *total correlation*

$$h(r) = g_{(2)}(r) - 1 \qquad (3.47)$$

is simply the fluctuation about the uniform probability $g_{(2)} = 1.00$ of finding a second particle located a distance r from another located at the origin. The breakdown of the total correlation into its direct and indirect components is made through the Ornstein–Zernike equation, first proposed in 1914 in the discussion of critical fluctuations:

$$h(12) = c(12) + \rho \int h(23)c(13) \, d3$$

$$\text{total} \qquad \text{direct} \qquad\qquad \text{indirect} \qquad\qquad\qquad (3.48)$$

This equation tells us nothing new—it is merely a defining relation for $c(12)$ and the solution of the structural problem is now transferred to the determination of the direct correlation $c(12)$. It is important to realize that $c(r)$ has no immediate physical interpretation for a pair of particles taken in *isolation*: it must relate to the pair in the presence of the remaining $(N - 2)$ particles.

According to (3.48), the total correlation between two atoms arises from a direct effect between 1 and 2, this modified indirectly through the average effect of a representative particle 3, subject to its remaining correlated with 2, $h(23)$.

If we try to interpret (3.48) in terms of its consequences for the potential of mean force $\Psi(12)$, the situation is not so clear as in (3.25). If we immerse a particle 1 in a *uniform* molecular system ($g_{(2)} = 1\cdot00$ initially) the local distribution will be perturbed under the action of the direct and indirect potentials $\Phi(12)$ and $W(12)$. We may therefore consider a *screened* or effective direct potential operating in the fluid, and the quantity $-kTc(12)$ bears this simple interpretation. At low densities we would, of course, expect $-kTc(12)$ to reduce to its unscreened form $\Phi(12)$, and this we may show as follows.

From (3.48),

$$h(12) \rightarrow c(12) \quad \text{as } \rho \rightarrow 0$$

i.e.

$$g_{(2)}(12) \rightarrow 1 + c(12)$$

Under these conditions $\Psi(12) \rightarrow \Phi(12)$, i.e.

$$g_{(2)}(12) = \exp\left(-\frac{\Psi(12)}{kT}\right) \rightarrow \exp\left(-\frac{\Phi(12)}{kT}\right) \rightarrow 1 - c(12)$$

whereupon

$$c(12) \sim \exp\left(-\frac{\Phi(12)}{kT}\right) - 1 \tag{3.49}$$

and

$$-kTc(12) \sim \Phi(12) \quad \text{when } kT \gg \Phi(12)$$

In spite of its somewhat obscure physical interpretation, the direct correlation function $c(r)$ is important in the theory of fluids in two respects. Firstly, it is directly accessible experimentally and the same scattering data can be used to determine both $g_{(2)}(r)$ and $c(r)$. Secondly, the range of the direct correlation function is approximately that of the pair potential to which it is relatively simply related. Certain limiting relations between the long-range form of $\Phi(r)$ and $c(r)$ exist, and these are extremely important in the discussion of the controversial long-range form of the pair potentials. Of immense importance is the fact that $c(r)$, having a range $\sim \sigma$, requires for its experimental determination scattering data down to angles $k \sim 2\pi/5\sigma$, whilst for the long-range function $g_{(2)}(r)$ scattering is difficult to separate from the 'straight-through' beam and in consequence long-range functions are rather imprecisely defined.

At low densities $c(r)$ is seen to reduce to the Mayer f-function (3.49), and it is clear both from this and from Figure 3.15 that the direct correlation is responsible for the short-range structure.

A knowledge of the direct correlation inserted in the Ornstein–Zernike relation (3.48) would enable us to make an analytic determination of the structure.

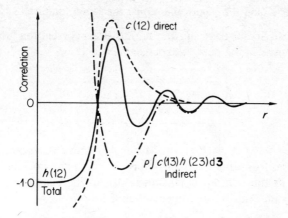

Figure 3.15 Qualitative breakdown of the total correlation into its direct and indirect components

Two important approximate forms for $c(r)$ have been proposed and these are most conveniently discussed in terms of a diagrammatic density expansion. Not surprisingly, the leading term in the diagrammatic series is the Mayer f-function, which dominates at low densities.

The diagrammatic representation expresses the total correlation $h(12)$ in terms of indirect effects through all possible chains of direct correlation between particles 1 and 2. This may be seen by repeatedly replacing $h(r)$ by $[c(r) + \text{integral}]$ within the integral of (3.48):

$$h(12) = c(12) + \rho \int c(13)c(23)\,\mathrm{d}3 + \rho^2 \iint c(13)c(34)c(42)\,\mathrm{d}3\,\mathrm{d}4 + \ldots$$

$$= \bullet\!-\!\!\bullet + \rho \; \overset{\circ}{\underset{\bullet\ \bullet}{\wedge}} + \rho^2 \; \overset{\circ\cdots\circ}{\underset{\bullet\ \ \bullet}{} } + \ldots \tag{3.50}$$

where ...... represents the direct correlation $c(ij)$.

The series representation of the total correlation is seen to be simple. If we knew the analytic form of the c-bond, and if we could write (3.50) in closed form, i.e. sum the series, we should know the structure. The two main first-order approximations to the direct correlation function consist of limited sums over the cluster-diagrammatic representations of $c(r)$. Neither sum is diagrammatically complete, but the physical implications are reasonably clear and enable $c(r)$ to be expressed in closed form. The hypernetted chain (HNC) approximation evolved through a series of intermediate stages, and the name is indicative of the types of diagrams retained in the approximation. The Percus–Yevick (PY) approximation was not developed on the basis of cluster techniques at all, but has been subsequently shown by Stell to have a simple diagrammatic representation and we shall discuss it in these terms.

The HNC and PY Approximations: the Diagrammatic Approach

Rushbrooke and Scoins have shown that the direct correlation function may be expressed in terms of the density expansion

$$c(12) = \sum_{n \geqslant 1} \alpha_{n+1}(12)\rho^{n-1} \tag{3.51}$$

where $\alpha_2(12) = f(12)$, i.e. $\alpha_2(12)$ is the Mayer f-function. As we anticipated in the last section, and from the low-density form of the direct correlation (3.49), the leading term in the density expansion is $f(12)$. The coefficients $\alpha_{n+1}(12)$ are given formally in terms of the Mayer f-bonds as

$$\alpha_{n+1}(12) = \frac{1}{(n-1)!} \int \ldots \int \sum \prod f(ij)\, d3 \ldots d(N-1)$$

so that, for example

$$\alpha_3(12) = \int f(12)f(23)f(13)\, d3$$

and

$$
\begin{aligned}
\alpha_4(12) = \frac{1}{2}\Bigg[\int\int \{ &f(12)f(23)f(34)f(14)f(24)f(13) \\
&+ f(12)f(23)f(34)f(14)f(24) + f(12)f(23)f(34)f(14)f(13) \\
&+ f(12)f(23)f(34)f(24)f(13) + f(12)f(14)f(34)f(24)f(13) \\
&+ f(12)f(23)f(14)f(24)f(13) + f(23)f(34)f(14)f(24)f(13) \\
&+ f(12)f(23)f(34)f(14) + f(12)f(13)f(34)f(24) \\
&+ f(13)f(23)f(24)f(14) \} \, d3\, d4 \Bigg]
\end{aligned}
$$

The coefficient $\alpha_4(12)$ exhibits a number of important features:

(i) The second, third, fourth and fifth terms in $\alpha_4(12)$ are *numerically* and *topologically* identical. These four terms are represented graphically as 4 .

The eighth and ninth terms are similarly identical, and are represented graphically as 2 .

(ii) Eight of the ten terms appearing in $\alpha_4(12)$ contain the factor $f(12)$, and these diagrams obviously decay rapidly with distance, having the range of the pair potential.

(iii) The two terms , having no $f(12)$ bond obviously have a range greater than the pair potential. For discontinuous potentials such as the hard-sphere interaction, those graphs which do not possess the $f(12)$ bond ensure the continuity of $c(12)$ even though $\Phi(12)$ is itself discontinuous.

The first few terms in the Rushbrooke–Scoins expansion of the direct correlation (3.51) may therefore be expressed diagrammatically:

$$
c(12) = \bullet\!\!-\!\!\bullet + \rho[\;\triangle\;]
$$

$$
+ \frac{\rho^2}{2}[2\;\square + 4\;\boxtimes + \boxtimes + \boxtimes + \boxtimes + \boxtimes\;] + \ldots \qquad (3.52)
$$

topologically identical diagrams are weighted in accordance with (i), above.

As usual, the root or base points remain at a fixed separation r_{12} whilst the field points ○ range over all space, subject to the constraining f-bonds.

The *individual* diagrams arising in (3.52) have little physical significance: the particular diagrams which develop do so as a consequence of our choice of the pair potential approximation. Groups of diagrams of the same order in density *collectively* represent the n-particle contribution to the direct correlation, but there is not the immediate physical interpretation that was available in the case of the virial expansion (2.40), for example.

$\alpha_5(12)$ contains 238 terms, 166 of which contain the $f(12)$ factor.

A diagrammatic expansion such as (3.52) makes little progress towards a closed expression which can be inserted in the Ornstein–Zernike equation and solved for $h(12)(= g(12) - 1)$. Various *classes of diagram* may be identified in the Rushbrooke–Scoins expansion and certain of these classes may be summed into closed expressions however. The HNC and PY approximations retain only certain classes of diagrams—those which can be conveniently summed. The remainder are neglected in the first-order theory.

The Diagrammatic Classes

The diagrammatic approach to the HNC and PY approximations consists of identifying various classes of diagram, and then expressing the correlation functions $c(r)$, $h(r)$, etc. as linear combinations of the classes.

The main classes of diagram and their characteristics are listed in Table 3.3. No distinction will generally be made between the simple and netted chain diagrams: $C(r)$ represents the entire sum over both kinds of chain diagram. A distinction is necessarily drawn between the $B(r)$ and $B'(r)$ diagrams, even though the latter is a subset of $B(r)$. Again, $B(r)$ represents the sum over the entire bundle class (including $B'(r)$), whilst $B'(r)$ represents the sum over those bundle diagrams without the $f(12)$ bond. The sum over the remaining diagrams is $E(r)$—the so-called elementary diagrams. In fact they are far from elementary to evaluate, and are usually dropped at the first opportunity.

78

Table 3.3

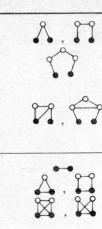	Chains $C(r)$:[a] clusters with at least one nodal field point. *Simple chain* is a cluster in which every field point (○) is a node—cutting at a node would cause the diagram to fall into two parts, each having a root or base point (●).
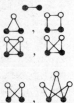	*Netted chains* are formed from simple chains by adding not more than one field point across each simple chain link. No chain diagram has a direct 12 link: inclusion of such a bond would convert the cluster into a *bundle*.
	Bundles $B(r)$:[b] clusters containing parallel collections of links between the two root points. There are always at least two *independent* routes from one root point to the other, one of which may be the direct 12 link. ●—● is arbitrarily included as the first bundle diagram.
	$B'(r)$ clusters: a subset of $B(r)$ diagrams having no 12 link.
	Elementary clusters, $E(r)$:[c] those diagrams which are neither chains nor bundles. An elementary graph cannot have a 12 bond since that would turn it into a bundle.

[a] Also termed *nodal diagrams*.
[b] Also termed *parallel diagrams*.
[c] Also termed *bridge diagrams*.

Diagrammatic Representation of the Correlations

The correlation functions may be expressed directly in terms of linear combinations of the classes $C(r)$, $B(r)$ and $E(r)$.

The total correlation $h(r)$ is the sum of all connected graphs through all possible modes of connection

$$h(r) = C(r) + B(r) + E(r) \tag{3.53}$$

From the Rushbrooke–Scoins expansion of the direct correlation $c(r)$ (3.51, 3.52) we see that no nodes arise, whereupon we drop the nodal class of chain diagrams $C(r)$:

$$c(r) = B(r) + E(r) \tag{3.54}$$

Finally, it may be shown that the indirect potential supplement $W(r) = \Psi(r) - \Phi(r)$ may be written as a sum over diagrams which exclude direct 12 interactions:

$$\frac{\Phi(r) - \Psi(r)}{kT} = C(r) + E(r) \tag{3.55}$$

We immediately have an exact closed analytic expression for the direct correlation, for from equations (3.53), (3.54) and (3.55)

$$c(r) = g_{(2)}(r) - 1 - \ln g_{(2)}(r) - \frac{\Phi(r)}{kT} + E(r) \tag{3.56}$$

Equation (3.56) represents the exact expression of the direct correlation, and its insertion in the Ornstein–Zernike equation (3.48) would, in principle, yield the exact total correlation: the main difficulty lies in the analytic expression of the elementary class, $E(r)$ for which there is no closed expression as yet.

Equation (3.54) may be rewritten as follows

$$c = B + E \tag{3.57}$$

$$= f_{12}(1 + C + B' + E) + B' + E \tag{3.58}$$

This is so since, from Table 3.3, the inclusion of the direct $f(12)$ bond as a factor in the C, B' and E diagrams converts them into B diagrams:

$$f(12)C \rightarrow B, \qquad f(12)E \rightarrow B, \qquad f(12)B' \rightarrow B$$

Equation (3.58) may be written as a combination of long-range and short-range components, remembering the range of the f and $(1 + f)$ functions (Figure 3.12):

$$\boxed{c = f(1 + C) + (1 + f)(B' + E)} \tag{3.59}$$

In the case of the hard-sphere interaction for which $f(12)$ is a negative step function, the short-range core and the long-range tail are quite distinct regions of the correlation function. For more realistic interactions the resolution of the long- and short-range components is not so clear. Nevertheless, (3.59) is in a convenient form for the discussion of the two main approximations.

The Hypernetted Chain (HNC) Approximation

The HNC approximation consists of dropping the (difficult) elementary class of diagrams, $E(r)$. This amounts to assuming a direct correlation function of the form

$$c_{\text{HNC}}(r) = g_{(2)}(r) - 1 - \ln g_{(2)}(r) - \frac{\Phi(r)}{kT} \tag{3.60}$$

(from (3.56)) which, when inserted in the Ornstein–Zernike equation yields the total correlation $h_{\text{HNC}}(r)$. In terms of the classes retained (3.57) becomes

$$c_{\text{HNC}} = B$$

or in long- and short-range form:

$$c_{\text{HNC}} = f(1 + C + E) + (1 + f)B' \tag{3.61}$$

which is seen to differ from the exact expression (3.59) both in the core and in the tail. In as far as the direct correlation contributes significantly to the short-range form of the total correlation (Figure 3.15), we may anticipate an inexact specification of the principal peak of the radial distribution upon which virtually all the thermodynamic quantities depend (Figure 3.16). The pressure equation, or equation of state (3.22) is particularly susceptible to small errors in the amplitude

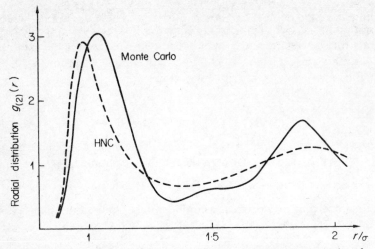

Figure 3.16 Discrepancy between the HNC and simulated principal peak of the liquid argon RDF

and/or position of the first peak of the RDF: Gaskell has exposed serious defects in the hyperchain theory when used to calculate fluid pressure.

At the zeros in the total correlation and as $r_{12} \to \infty$, the following condition applies, from (3.60):

$$\frac{\Phi(12)}{kT} = -c_{HNC}(12)$$

An experimental measurement of the direct correlation provides direct access according to the HNC approximation, to the pair potential operating in the fluid and this expression has been used to establish the controversial long-range forms of $\Phi(12)$.

The Percus–Yevick (PY) Approximation

This approximation consists of dropping the long-range component $(1 + f)$ $(B' + E)$ from equation (3.59), but preserving the *exact* short-range form:

$$c_{PY} = f(1 + C) \tag{3.62}$$

Algebraic manipulation of (3.53), (3.54) and (3.55) yields the following analytical expression

$$c(r) = h(r) - \ln\left[1 + (y(r) - 1)\right]$$

where

$$y(r) = g_{(2)}(r) \exp\left(\frac{\Phi(r)}{kT}\right)$$

If we now linearize the logarithm, and introduce a new class of diagrams $D(r)$ which preserves the exact relation (3.57) we have

$$c(r) = h(r) - [y(r) - 1] + D(r)$$

and then set $D(r) = 0$, we obtain

$$c_{PY}(r) = h(r) - \left\{ g_{(2)}(r) \exp\left(\frac{\Phi(r)}{kT}\right) - 1 \right\} \tag{3.63}$$

which may be inserted in the Ornstein–Zernike equation (3.48), and solved for the total correlation.

In as far as the PY approximation preserves the exact short-range form of the direct correlation (3.62), we might anticipate that for short-range interactions the Percus–Yevick approximation should give a superior representation of the short-range structure of the total correlation, and this is seen to be the case from Figure 3.17.

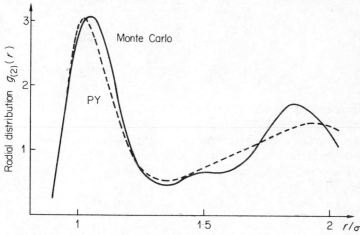

Figure 3.17   Discrepancy between the PY and simulated principal peak of the liquid-argon RDF

A zero in the total correlation implies

$$\frac{\Phi(12)}{kT} = \ln\left[1 - c_{PY}(12)\right]$$

and this also applies asymptotically as $r_{12} \to \infty$. This provides direct experimental access through a measurement of $c(r)$ to the long-range form of the pair potential, although there is only asymptotic agreement with the HNC estimate.

Comparison of the HNC and PY Approximations

It is clear from (3.61) and (3.62) that PY represents the more drastic approximation in terms of the number and classes of diagrams dropped (Table 3.4).

82

Table 3.4

Approximation	Diagrams retained	Diagrams dropped
HNC	B	E
PY	$f(1 + C)$	$(1 + f)(B' + E)$
Exact	$f(1 + C) + (1 + f)(B' + E)$	

Diagrammatically, in the Rushbrooke–Scoins expansion (3.52)

$$c(r) = \;\bullet\!\!-\!\!\bullet\; + \rho[\triangle]$$

$$+ \frac{\rho^2}{2}[2\,\square + 4\,\boxslash + \boxtimes + \boxtimes + \boxtimes + \boxtimes] + \cdots$$

the third term first shows the signs of approximation, becoming

$$\frac{\rho^2}{2}[2\,\square + 4\,\boxslash + \boxtimes + \boxtimes + \boxtimes] \qquad \text{HNC} \quad (3.64)$$

$$\frac{\rho^2}{2}[2\,\square + 4\,\boxslash] \qquad\qquad\qquad \text{PY} \quad (3.65)$$

where the diagrams retained and dropped in the two cases are in accordance with Table 3.4.

We have already noted that in preserving the exact short-range form of the direct correlation function the PY approximation gives a superior structural and thermodynamic representation of a dense fluid, in particular for short-ranged potentials.

The diagrams dropped in the two approximations are

$$\boxtimes \qquad\qquad \text{HNC}$$

$$\boxtimes + \boxtimes + \boxtimes + \boxtimes \qquad \text{PY}$$

and the apparently paradoxical superiority of the PY approximation can be understood from a comparison of the pairs of diagrams $\boxtimes$ and $\boxtimes$. The field point accessibility of the two diagrams is almost identical, and for the hard-sphere interaction the two diagrams differ virtually only in sign. This cancellation has been confirmed by Klein for the long-range diagrams $\boxtimes$, $\boxtimes$ (Figure 3.18), and of course, applies equally to the short-range diagrams $\boxtimes$ and $\boxtimes$.

Thus the neglect of four diagrams in the PY case in fact represents a less drastic approximation in terms of self-cancellation amongst the diagrams than the HNC case.

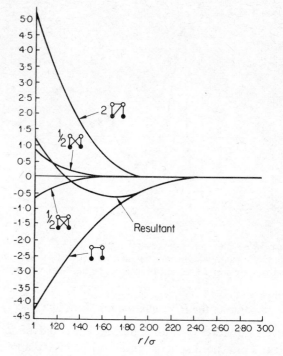

Figure 3.18 Hard-sphere cluster integrals involved in the first-order HNC and PY approximations

Of course, for more realistic interactions the cancellation cannot be discussed in such simple terms, for then the Mayer f-function has a more complicated temperature-dependent form. Nevertheless, when the repulsive forces dominate we can expect the PY approximation to give a better representation of the structural features of a dense fluid.

Ashcroft and Lekner have compared the experimental structure factor $S(k)$ of several liquid metals to the PY hard-sphere calculation in which it is assumed that the ion–ion interaction is adequately described by the hard-sphere potential. A comparison of the calculated PY structure factor at a packing fraction $\pi\rho\sigma^3/6 = 0.45$ is in essential agreement with the experimental data for rubidium at 40 °C. The essentially geometric packing problem is well described by the hard-sphere model as we see from the quantitative agreement in the vicinity of the principal peak, and it is this feature of the structure factor upon which electrical resistivity calculations are sensitive. This indicates that the 'wavelength' of the total correlation is correct, but the progressive discrepancy with increasing k suggests that the details of the model interaction are not quite correct (Figure 3.19).

84

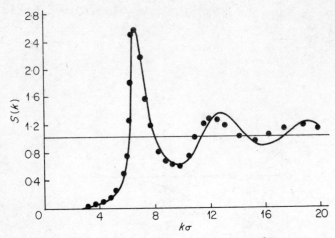

Figure 3.19 The hard-sphere and liquid-Rb (40 °C) structure factor in the vicinity of the melting point. The principal structural features are seen to be reproduced by the hard-sphere curve

Solution of the PY Equation for Hard Spheres

The Percus–Yevick equation has the quite unexpected advantage that it may be solved analytically for hard spheres. Insertion of the PY approximation for the direct correlation (3.63) in the Ornstein–Zernike equation yields

$$g_{(2)}(12) - c(12) = 1 - \rho \int c(13)\,d3 + \rho \int g_{(2)}(23)c(13)\,d3 \qquad (3.66)$$

Now, although both $g_{(2)}$ and c have finite discontinuities at the hard core surface, the quantity $(g - c)$ will be *continuous* at $r_{12} = \sigma$ together with its first D derivatives for dimension D. This we may see diagrammatically, for

$$g_{(2)}(12) - c(12) = (1 + C + B + E) - (B + E)$$
$$= (1 + C) \qquad (3.67)$$

and the chain diagrams C, having no direct $f(12)$ bond, are evidently continuous at σ. In the PY approximation we have

$$\left.\begin{array}{ll} g_{(2)}(12) = 0 & r \leqslant \sigma \\ c(12) = 0 & r > \sigma \end{array}\right\} \qquad (3.68)$$

We have to determine the short-range form ($r_{12} \leqslant \sigma$) of the direct correlation, and we therefore investigate (3.66) in the vicinity of $r_{12} = 0$. From (3.68) it is clear that the second integral in (3.66) will only contribute in the vicinity of $r_{13} \sim \sigma$ (Figure 3.20) since only in the shaded region are both functions $g(23)$ and $c(13)$ non-zero.

We may form a Taylor expansion about $r = 0$ therefore, expanding the integral about $r_{13} = \sigma$ since this is the only region which contributes. Before

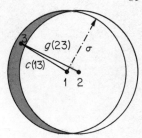

Figure 3.20   Geometry for the analytic solution of the PY equation

expanding we may rewrite the second integral from the general relation:

$$\int g_{(2)}(23)c(13)\,d3 = \frac{2\pi}{r_{12}}\iint_{\text{strip}} tg_{(2)}(t)c(13)\,d3\,dt. \tag{3.69}$$

where integration is over the strip $|r_{13} - t| \leqslant r_{12} \leqslant |r_{13} + t|, r_{13} \geqslant 0, t \geqslant 0$. We then obtain from (3.66) for $c(12)$

$$
\begin{aligned}
- c(12) = 1 &- \rho \int c(13)\,d3 + 2\pi\rho\left\{\frac{1}{2!}r_{12}[\sigma g(\sigma)\sigma\rho(\sigma)]\right. \\
&+ \frac{r_{12}^2}{3!}[\sigma g(\sigma)\sigma c(\sigma)]' + \frac{r_{12}^3}{4!}[\sigma g(\sigma)\sigma c(\sigma)]'' \\
&\left.+ \frac{r_{12}^4}{5!}[\sigma g(\sigma)\sigma c(\sigma)]''' + \ldots\right\}
\end{aligned}
\tag{3.70}
$$

where the second integrand in (3.66) has been rewritten according to (3.69) and Taylor expanded about $r_{13} = \sigma$.

Equation (3.70) is a polynomial in r_{12}: we have yet to determine the coefficients. This we may do by differentiating at the origin, and using the equality of derivatives of g and $-c$ at $r_{12} = \sigma$ (this follows from the continuity of $g - c$ at σ, together with (3.68)):

$$
\left.
\begin{aligned}
-c(0) &= 1 - \rho \int c(13)\,d3 \\
-c'(0) &= -\pi\rho\sigma^2 c(\sigma)g(\sigma) = -\pi\rho\sigma^2 c(\sigma)^2 \\
-c''(0) &= 0 \qquad\qquad\qquad \text{(equality of derivatives)} \\
-c'''(0) &= \tfrac{1}{12}\pi\rho\{\sigma^2(c'(\sigma)^2 - 2c(\sigma)c''(\sigma)) + 2\sigma c(\sigma)c'(\sigma) + c(\sigma)^2\} \\
-c''''(0) &= 0
\end{aligned}
\right\}
\tag{3.71}
$$

These results imply a polynomial of the form:

$$c_{\mathrm{PY}}(r) = -\frac{(1 + 2\eta)^2}{(1 - \eta)^4} + \frac{6\eta(1 + \tfrac{1}{2}\eta)^2}{(1 - \eta)^4}\frac{r}{\sigma} - \frac{\eta(1 + 2\eta)^2}{2(1 - \eta)^4}\left(\frac{r}{\sigma}\right)^3 \tag{3.72}$$

where $\eta = \tfrac{1}{6}\pi\rho\sigma^3$, $r < \sigma$. We see both from (3.72) and Figure 3.21 that $c_{\mathrm{PY}}(r)$ recovers its negative-step-function form in the limit $\rho \to 0$.

86

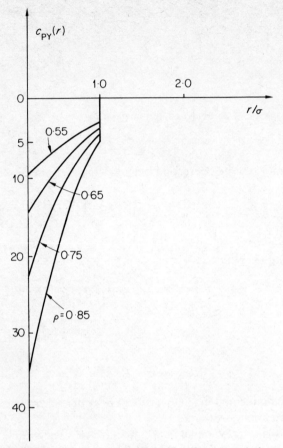

Figure 3.21 The PY hard-sphere direct correlation
function (cf. Figure 5.18)

The PY hard-sphere pressure and compressibility equations of state follow
directly from (3.20) and (3.22)

$$\left(\frac{PV}{NkT}\right)_p = \frac{1 + 2\eta + 3\eta^2}{(1 - \eta)^2} = \frac{1 + \eta + \eta^2 - 3\eta^3}{(1 - \eta)^3} \qquad (3.73)$$

$$\left(\frac{PV}{NkT}\right)_c = \frac{1 + \eta + \eta^2}{(1 - \eta)^3} \qquad (3.74)$$

and these are seen to be inconsistent over the entire density range, though
particularly when

$$3\eta^3 \geqslant 1 + \eta + \eta^2$$

which in fact limits η to ~ 0.3 after which the discrepancy becomes significant. cant. In fact, the two pressure relations appear to bracket the correct isotherm—
the precise numerical consequences will be discussed in Chapter 4.

Both equations yield monotonic isotherms as a function of density, the only singularity occurring at $\eta = 1$ which represents a physically impossible density, and certainly cannot be identified with the solid–fluid phase transition. For relative densities $\rho/\rho_0 \gtrsim 0.8$, $g_{(2)}(r)$ takes on negative values, and so at these high densities the PY solutions can be rejected on physical grounds. Even this is an impossibly high density however, since the relative density cannot exceed that of a regular close-packed array of spheres, 0.7405.

Temperley has observed that there are solutions to the PY equation having the form $\cos a_i r$ where the coefficients a_i are reciprocal distances. These oscillatory solutions, moreover, resemble the form of the distribution in a solid. Hutchinson has since shown that unfortunately these solutions imply negative intensity of scattered radiation at certain scattering angles, and are therefore physically unacceptable. The scattering intensity criterion is unduly harsh and allows little scope for approximation: we shall return to this point in our discussion of phase transitions (Chapter 5).

Structure Factor and the Ornstein–Zernike Equation

The structure factor $S(k)$ is simply related to the Fourier transform of the total correlation:

$$S(k) = 1 + 4\pi\rho \int_0^\infty [g_{(2)}(r) - 1]\frac{\sin kr}{kr}r^2\,dr$$

$$= 1 + 4\pi\rho \int_0^\infty h(r)\frac{\sin kr}{kr}r^2\,dr = 1 + \rho h(k) \tag{3.75}$$

The Ornstein–Zernike equation (3.48) may be Fourier transformed with the aid of the convolution theorem to give

$$h(k) = c(k) + \rho c(k)h(k) \tag{3.76}$$

so that

$$c(k) = \frac{h(k)}{1 + \rho h(k)} = \frac{S(k) - 1}{\rho S(k)} \tag{3.77}$$

from which we see that the direct correlation is experimentally accessible through the scattering function $S(k)$. Alternatively we may express $S(k)$ in terms of the direct correlation:

$$S(k) = \frac{1}{1 - \rho c(k)} \quad \text{or} \quad \rho c(k) = 1 - \frac{1}{S(k)} \tag{3.78}$$

Thus, a calculation of $c(k)$ is readily converted into $S(k)$.

A condition on $S(k)$, other than that it must, of course, be positive, is provided by a *sum rule* which follows directly from the inverted form of (3.75)

$$h(r) = \frac{1}{2\pi^2\rho} \int_0^\infty [S(k) - 1]\frac{\sin kr}{kr}k^2\,dk$$

which at $r = 0$ reduces to

$$2\pi^2\rho = \int_0^\infty [S(k) - 1]k^2 \, dk \qquad (3.79)$$

and is a restatement of the normalization condition (1.9). This is not a stringent constraint on the detailed form of $S(k)$ but it does nevertheless provide a criterion of acceptability for model structure factors.

There is also a well known thermodynamic relation between the long-wavelength limit $(k \to 0)$ of the structure factor $S(0)$ and the isothermal compressibility, χ_T. From (3.22) we have

$$\frac{1}{\chi_T} \cdot \frac{1}{kT} = 1 - 4\pi\rho \int_0^\infty c(r)r^2 \, dr \qquad (3.80)$$

where the integral represents the $k = 0$ form of $c(k)$:

$$\rho c(k) = 4\pi\rho \int_0^\infty c(r)\frac{\sin kr}{kr}r^2 \, dr \qquad (3.81)$$

The compressibility relation may then be rewritten with the aid of (3.78)

$$\boxed{S(0) = kT\rho\chi_T}$$

As the density of the system rises so its compressibility will decrease and $S(0) \to 0$. In the vicinity of the critical point and the fluid–solid phase transition where the compressibility is large, we would expect an anomalous rise in the small-angle scattering, and this is observed to be the case (Figure 3.22). We may understand this physically as the development of long-wavelength cooperative isothermal density fluctuations in the liquid. In principle a measurement of $S(0)$ is equivalent to the determination of the equation of state: in practice small-angle scattering measurements are exceedingly difficult.

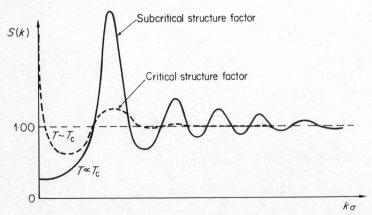

Figure 3.22 The structure factor $S(k)$ in the vicinity of critical point. Note the long-wavelength development $(k \to 0)$

Consequences of the Direct Correlation: the Empty-Core Model

Whilst the Percus–Yevick theory of the direct correlation yields relatively simple expressions for $c(r)$ and the equations of state for a hard-sphere fluid, we may exploit the insensitivity of $c(k)$ to the short-range form of the direct correlation function (3.81) to discuss a simple model in which all the principal thermodynamic and structural features may be easily evaluated.

We approximate $c_{PY}(r, \eta)$ by a rectangular 'empty-core' function of depth $c_{PY}(\sigma, \eta) = \{(\eta(3 - \eta^2) - 2)/2(1 - \eta)^4\}$ (Figure 3.23a). The Fourier transform (3.81) is simple to evaluate:

$$c_{\text{empty core}}(k) = 12\eta \left\{ \frac{\eta(3 - \eta^2) - 2}{(1 - \eta)^4} \right\} \left\{ \frac{\sin k\sigma - k\sigma \cos k\sigma}{k^3 \sigma^3} \right\}$$

$$= 12\eta \left\{ \frac{\eta(3 - \eta^2) - 2}{(1 - \eta)^4} \right\} j_1(k\sigma)/k\sigma \tag{3.82}$$

(where j_1 is the first-order spherical Bessel function) which has the long-wavelength limit

$$c_{\text{empty core}}(k) = 4\eta \left\{ \frac{\eta(3 - \eta^2) - 2}{(1 - \eta)^4} \right\} \tag{3.83}$$

The structure factor follows as

$$\frac{1}{S(k)} = 1 - 12\eta \left\{ \frac{\eta(3 - \eta^2) - 2}{(1 - \eta)^4} \right\} \left\{ \frac{\sin k\sigma - k\sigma \cos k\sigma}{k^3 \sigma^3} \right\}$$

$$= 1 - 12 \left\{ \frac{\eta(3 - \eta^2) - 2}{(1 - \eta)^4} \right\} j_1(k\sigma)/k\sigma \tag{3.84}$$

which is seen to show the correct qualitative features (Figure 3.24): $S(k)$ has a damped oscillatory form, the peaks becoming sharper with increasing density, although the small progressive shift of $S(k)$ to higher wavenumbers with η is not given in this model. The first and principal peak located at $k \sim 2\pi/\sigma$ corresponds to an oscillation in $g_{(2)}(r)$ of wavelength $\sim \sigma$ indicating the essentially geometric aspect of hard-sphere packing in a dense fluid. The subsequent peaks also develop with increasing density and this is indicative of the development of fine structure in the radial distribution function. The long-wavelength limit to (3.84) is easily determined, the trigonometric factor $\rightarrow \frac{1}{3}$ as $k \rightarrow 0$:

$$S(0) = \left[1 - 4\eta \left\{ \frac{\eta(3 - \eta^2) - 2}{(1 - \eta)^4} \right\} \right]^{-1} \tag{3.85}$$

The packing fraction in the vicinity of the melting point is $\eta \sim 0.46$ and (3.85) yields $S(0) = 0.0673$ which is seen to be significantly higher than the values quoted in Table 3.5 for a number of simple liquids. This is not surprising since the empty-core model underestimates the exclusion characteristics of $c(r)$ in the

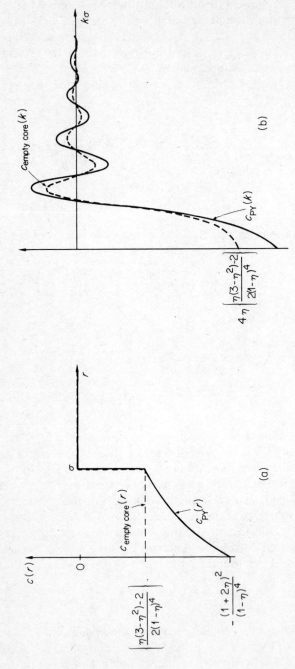

Figure 3.23 (a) Hard-sphere direct correlation in the PY and 'empty-core' approximations. (b) The transforms $c(k)$ of the PY and empty-core direct correlations

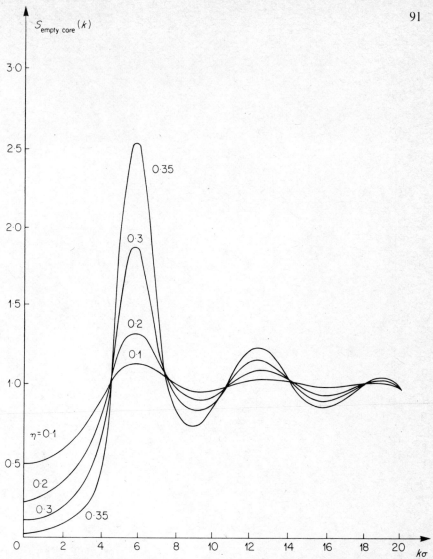

Figure 3.24 Variation of structure factor with density in the empty-core model

Table 3.5. $S(0)$ at the melting point for simple liquids (Faber)

Ar	Li	Na	K	Rb	Cs	Cu
0·064	0·031	0·024	0·023	0·022	0·028	0·021
Ag	Zn	Cd	Hg	Al	Ga	In
0·019	0·014	0·012	0·005	0·018	0·005	0·007
Tl	Sn	Pb	Sb	Bi		
0·011	0·008	0·009	0·020	0·011		

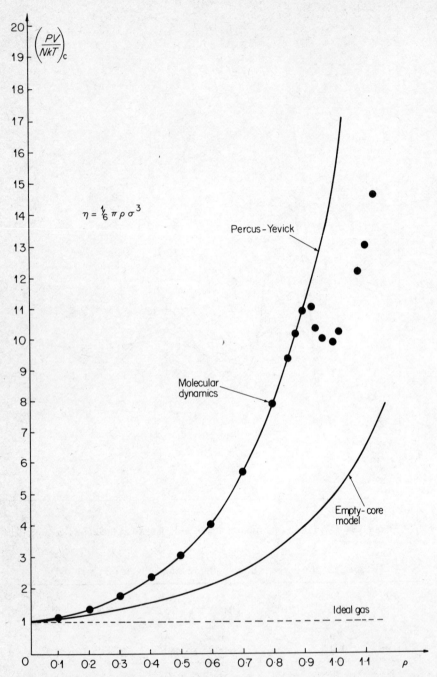

Figure 3.25 Comparison of hard-sphere equations of state with the molecular dynamics results

core region: the simple rectangular core assumed here presumably is more representative of a 'soft' repulsive interaction and to a certain extent this is supported by the result for argon.

The equation of state follows from the compressibility relation (3.80) as

$$\left(\frac{PV}{NkT}\right)_c = \frac{7 + 5\eta(2 - \eta)}{(1 - \eta)} + 16 \ln(1 - \eta) - 6 \qquad (3.86)$$

and the isotherm is compared with the hard-sphere PY equation of state in Figure 3.25. The gradient of the empty-core isotherm at each density point is much lower than either the PY or molecular dynamics curves. This overestimate of the compressibility is to be directly attributed to the softness in the repulsive potential introduced in approximating the core of the direct correlation to a rectangular function. Analytically the disparity between the empty-core approximation and the 'true' PY core of the direct correlation increases with the packing fraction η. Of course, physically this is just what we should expect: the decrease in free volume with η means, for hard spheres, that the gradient must rapidly increase becoming infinite in the close-packed configuration.

There is no indication of a solid–fluid phase transition for either isotherm, although the molecular dynamics results clearly indicate such a phase change. A more detailed discussion will be postponed until Chapter 5.

Realistic Systems

In as far as the structure of a dense fluid is essentially a geometric packing problem, the repulsive core dominates the structural features of the fluid.. In fact a cohesive or attractive interaction must operate within the fluid and the results of the previous section, whilst illustrating the principal structural and thermodynamic features of a dense fluid, must be modified to incorporate the effects of the attractive branch of the pair potential.

The Rushbrooke–Scoins density expansion of the direct correlation is made in terms of the Mayer f-function which, for realistic low-temperature fluids, develops a weak positive long-range tail (Figure 3.26). With increasing temperature the realistic f-function tends to the rectangular function as the rigid-core aspect of the pair potential dominates. The consequences for the direct correlation are shown schematically in Figure 3.27. $c(r)$ is now, of course, both a function of density *and temperature*. It is convenient to split the compressibility equation of state into long- and short-range forms:

$$\frac{1}{kT}\left(\frac{\partial P}{\partial \rho}\right)_T = 1 - 4\pi\rho\left\{\underbrace{\int_0^\sigma c(r)r^2 \, dr}_{\text{core}} + \underbrace{\int_\sigma^\infty c(r)r^2 \, dr}_{\text{tail}}\right\} \qquad (3.87)$$

where the first and second integrals represent the contributions of the core and tail of the direct correlation, respectively.

At high temperatures the weak positive tail is washed out and the isotherm of a realistic assembly is hard-sphere like. The PY approximation would be

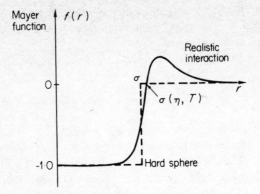

Figure 3.26 Comparison of the hard-sphere and realistic-Mayer f-function

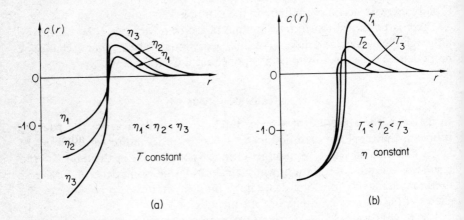

Figure 3.27 Schematic variation of a realistic direct correlation (a) with density (b) with temperature

expected to give a good description of the equation of state under these circumstances, as indeed it does (Figure 3.28). At lower temperatures the development of a positive tail in the direct correlation function partially *offsets* the core integral and lowers the pressure as shown in Figure 3.28 for argon.

From equation (3.87) we see that the equation of state is, in fact, highly sensitive to the form of the long-range tail, and it is at the same time evident that the PY approximation, in neglecting the long-range component of the direct correlation (3.62), can never provide an adequate description of realistic low-temperature, high-density systems.

The HNC approximation, on the other hand, whilst misrepresenting both the short- and long-range form of the direct correlation (3.61) nevertheless

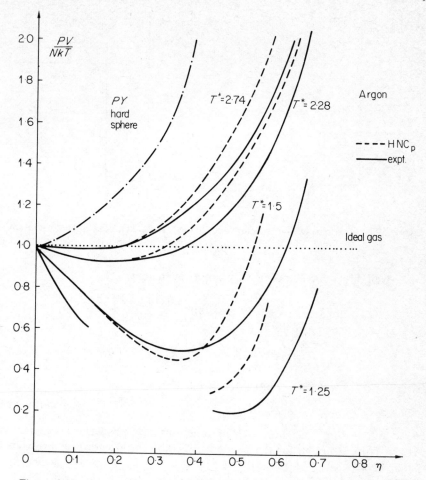

Figure 3.28 Comparison of the HNC and experimental argon isotherms. At high temperatures the isotherms approach the hard-sphere PY curve as the repulsive core dominates the interaction

makes some attempt at a long-range tail. Since the HNC core is incorrect it will never describe high-density, high-temperature systems in which the geometric features of hard-sphere packing dominate the statistical thermodynamics of the fluid. Nevertheless, Klein has concluded that at low temperatures and intermediate densities the HNC approximation is superior to PY for realistic systems, as a comparison in Figure 3.28 shows.

A simple model for a realistic system is provided by the square-well interaction (Figure 3.29a): the attractive region is of depth $-\varepsilon$ and range 2σ which is more or less representative of a realistic interaction. In Figure 3.29(b) the depth of the core of the direct correlation is $-D(\eta)$, whilst the height of the short-range tail is $H(\eta)$ and has the same range as the pair potential. The Fourier transform of the

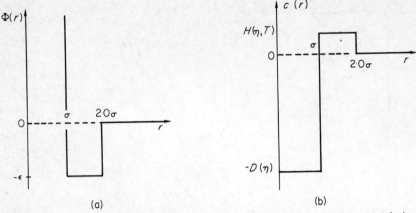

Figure 3.29 (a) Square-well interaction and (b) the associated model direct correlation, now a function of both temperature and density

square-well direct correlation is very simply evaluated:

$$c_{\text{square well}}(k) = -24\eta(D(\eta) + H(\eta))\left\{\frac{\sin(k\sigma) - k\sigma\cos(k\sigma)}{(k\sigma)^3}\right\}$$

$$+ 24\eta H(\eta)\left\{\frac{\sin(2k\sigma) - 2k\sigma\cos(2k\sigma)}{(2k\sigma)^3}\right\}$$

$$= -24\eta(D(\eta) + H(\eta))j_1(k\sigma)/k\sigma$$

$$+ 24\eta H(\eta)j_1(2k\sigma)/2k\sigma \tag{3.88}$$

The depth $-D(\eta)$ and height $H(\eta)$ functions are unspecified other than $|D(\eta)| > H(\eta)$, although for the purposes of comparison with the hard-sphere system we may set $-D(\eta) = \{(\eta(3 - \eta^2) - 2)/2(1 - \eta)^4\}$. The long-wavelength limit of (3.88) remains identical to the hard-sphere case:

$$c_{\text{square well}}(0) = -8\eta(D(\eta) + H(\eta)) + 8\eta H(\eta)$$

$$= -8\eta D(\eta)$$

$$= -4\eta\left\{\frac{\eta(3 - \eta^2) - 2}{(1 - \eta)^4}\right\} \tag{3.89}$$

(cf. equation (3.83)). Similarly, the long-wavelength limit to the structure factor:

$$S(0) = [1 - 8\eta D(\eta)]^{-1}$$

$$= \left[1 + 4\eta\left\{\frac{\eta(3 - \eta^2) - 2}{(1 - \eta)^4}\right\}\right]^{-1} \tag{3.90}$$

(cf. equation (3.85)). We may conclude that it is the repulsive component of the pair potential which dominates the compressibility of the fluid—but then this is intuitively obvious.

The first term in (3.88) serves to *enhance* the oscillations in $c_{\text{square well}}(k)$ with respect to the hard-sphere system, whilst the second more rapidly oscillating component, in antiphase to the first, rapidly damps out as $(R\sigma)^{-3}$ where R is the range of the attractive forces. It is clear that it is the principal peak in $c(k)$ which is most subject to modification although the effect is slight, and $S(k)$ does not differ significantly from system to system.

The compressibility equation of state is readily evaluated provided we assign a density and temperature dependence to the tail. Formally, the equation of state is simply (from (3.87)),

$$\left(\frac{PV}{NkT}\right)_c = 1 + 8 \int \eta D(\eta)\,d\eta - 40 \int \eta H(\eta, T)\,d\eta + c \qquad (3.91)$$

where c is a constant of integration chosen such that $(PV/NkT) \to 1$ as $\eta \to 0$. The first integral represents the rigid-core contribution to the isotherm and is essentially positive. The second represents the effect of the attractive forces, and at low temperatures accounts for the depression of the isotherm below the ideal-gas value (Figure 3.28). At high temperatures $H(\eta, T) \to 0$ over the entire density range and the isotherm tends to the hard-sphere curve.

An analogous analysis for the pressure equation of state yields similar isotherms, though much of the density and temperature dependence is included implicitly in the pair distribution $g_{(2)}(r, \eta, T)$.

Self-consistent Approximations (SCA)

Internal inconsistency between the pressure and compressibility equations of state arises as a consequence of approximation either in the form of the superposition approximation in the KBGY class of equations, or as incomplete diagrammatic summing in the case of the HNC and PY approximations. The pressure and compressibility isotherms tend to bracket the correct equation of state, the latter being on the high-pressure side for PY, and vice versa for HNC. Since the PY approximation accounts exactly for the core of the direct correlation function, whilst HNC includes a partial sum over the tail diagrams, some effort has gone into *forcing* self-consistency by forming a linear combination of c_{PY} and c_{HNC}. This device of enforcing thermodynamic consistency cannot be regarded as an advance of physical understanding although some diagrammatic justification for the combination can be made (Chapter 5); nevertheless, the results are excellent at least to the sixth virial coefficient for hard spheres. Nonetheless, it has in fact been shown that self-consistency is possible *only* for the exact distribution.

Rowlinson has proposed the self-consistent approximation (RSCA):

$$c_{\text{RSCA}}(r) = f(r)y(r) + \alpha(\rho, T)[y(r) - 1 - \ln y(r)] \qquad (3.92)$$

where $y(r) = \exp[\Phi(r)/kT]g_{(2)}(r)$. α is a pure number, dependent upon density and temperature, and Rowlinson proposes to represent $\alpha(\rho, T)$ as

$$\alpha(\rho, T) = \sum_{n=0}^{\infty} \rho^n \alpha_{n+4}(T)$$

Equation (3.92) reduces to the PY approximation for $\alpha = 0$ and to the HNC approximation for $\alpha = 1$. At low densities an optimum value of α may be determined by iteration which yields a self-consistent equation of state. Whilst such an equation tells us little of the physical processes operating in the liquid, an accurate and self-consistent equation of state is of considerable value in itself.

A similar relation, the generalized HNC (GHNC) approximation, has been proposed by Hurst

$$c_{GHNC}(r) = f(r)y(r) + y(r) - 1 - y^m(r) \ln y(r)$$

$$= c_{HNC}(r) + [1 - y^m(r)] \ln y(r) \tag{3.93}$$

m is again a simple numeric, and the expansion

$$m(\rho, T) = \sum_{n=0}^{\infty} \rho^n m_{n+4}(T)$$

has been proposed. Equation (3.93) reduces to the HNC case for $m = 0$, but cannot be reduced to PY form. Again, the value of such an approximation lies in its ability to yield an accurate and self-consistent equation of state rather than any physical insight it might provide at molecular level.

Functional Differentiation and Second-order Theories

The potential of mean force $\Psi(12)$ may be regarded as a linear combination of the pair potential $\Phi(12)$ and some unspecified supplement $W(12)$ representing the mean contribution of the remaining $(N - 2)$ particles:

$$\Psi(12) = \Phi(12) + W(12)$$

A knowledge of $W(12)$ would provide direct access to the pair distribution function through the relation $g_{(2)}(12) = \exp(-\Psi(12)/kT)$.

Obviously at low densities $\Psi(12) \to \Phi(12)$, the intervention of correlated $3, 4, \ldots$-body effects amongst the remaining $(N - 2)$ particles having negligible effect. This suggests that a density expansion for $W(12)$ might be appropriate, involving the averaged effects of $3, 4, \ldots$-body contributions as the density increases—these higher-order particles of course remaining correlated with particle 1. This scheme neglects both the functional dependence of the n-body contribution to $W(12)$ (i.e. through the pair potential function and interparticle correlations) and the density dependence of the contribution, for it is supposed to be a density expansion. The most general, non-commital mathematical statement we can make in the spirit of a density expansion is in terms of a *functional* Taylor expansion:

$$\ln g_{(2)}(12) + \frac{\Phi(12)}{kT} = \rho \int h(13)\frac{\delta \mathscr{F}}{\delta \xi}\,d3 + \frac{\rho^2}{2!}\int\int h(13)h(14)\frac{\delta^2 \mathscr{F}}{\delta \xi^2}\,d3\,d4 + \dots$$

$$(3.94)$$

This expansion truncated at the first integral bears considerable similarity to both the KBGY and the OZ(HNC–PY) classes of equations. Indeed, it may be shown that for suitable choices of the functional $\mathscr{F}$ and the function ξ we may regain the HNC, PY and BGY equations, although the choice of $\mathscr{F}$ and ξ appears quite arbitrary at the moment. Application of the method depends to a large extent upon the skill in choosing a 'successful' functional form: different equations result from different choices of $\mathscr{F}$ and ξ. A detailed consideration of the various forms for $\mathscr{F}$ and ξ, and the first-order theories that result, is given in *LSP*.

The rederivation of the first-order theories on the basis of (3.94) seems to place them on a sounder mathematical basis, but this is in many respects illusory. Both of the main classes of approximate equation have a form of which (3.94), terminated at the first term, is a general statement. There is, as we have observed, no rational basis for our choice of functionals $\mathscr{F}$ and ξ. The real importance of the functional Taylor expansion (3.94) lies in its ability to suggest second-order approximations through the term of order ρ^2. In principle a succession of higher-order terms could be included, but mathematical difficulties have so far restricted the expansion to the second term.

Retaining the same choices for $\mathscr{F}$ and ξ we may proceed directly to the second-order theories HNC2, PY2 and BGY2. For example, the direct correlation function in the HNC2 approximation becomes

$$c_{HNC2}(12) = c_{HNC1}(12) + \varepsilon(12) \tag{3.95}$$

which is to be compared with the exact relation

$$c(12) = c_{HNC1}(12) + E(12)$$

Equation (3.95) suggests that HNC2 takes some account of the elementary diagrams dropped in HNC1. Similarly for the PY2 approximation

$$c_{PY2}(12) = c_{PY1}(12) + \Delta(12) \tag{3.96}$$

where $\Delta(12)$ takes partial account of some of the diagrams $D(12)$ dropped in the PY1 approximation (equation (3.63)).

The resulting second-order integral equations are considerably more difficult to solve than their first-order counterparts, and published work is currently at the level of low-density solutions and the determination of successive virial coefficients.

This refinement of the first-order theories is not wholly satisfactory. Apart from the difficulties of obtaining liquid-density solutions, mathematical convenience has been obtained at the expense of a clear understanding of the physical and diagrammatic processes which the approximation represents.

Certainly the second-order theories represent extensions of the diagrammatic summations included in the first-order treatments, but as we have seen, a delicate cancellation scheme operates amongst the diagrams and more complete summings may actually *worsen* agreement at high densities. Again, whether the second-order theories will show any evidence of a solid–fluid phase transition is not yet known, but on the basis of our physical understanding of these partial diagrammatic extensions we have no grounds for anticipating a significant improvement on the first-order theories at intermediate and high densities, let alone a change of phase.

Perturbation Theories

The hard-sphere approximation provides an excellent basis for a perturbative extension to more realistic soft-sphere systems. The adequacy of the hard-sphere representation is shown in Ashcroft and Lekner's calculation of the structure factor at a reduced density of $\eta = 0.45$ (Figure 3.19). At small k-values the agreement with the experimental data for rubidium is excellent. At larger scattering angles, however, there is a progressive discrepancy, and this may be directly attributed to an inadequacy in the assumed form of pair potential. In as far as the structural features of dense fluids are dominated by the form of the repulsive branch of the pair interaction; the hard sphere represents a good first approximation. Of course, over some regions of the intermolecular separation the forces are attractive. Nevertheless, at high densities (and at high temperatures for all densities) the structure is dominated by the repulsive forces, and so an accurate theory of repulsive forces can provide a foundation for an equilibrium theory of fluids.

There have been several attempts to relate hard-sphere data to the properties of fluids with other repulsive potentials in terms of a perturbative expansion. It has been shown by molecular dynamics simulation that for fluids interacting through a $1/r^n$ potential the height of the first peak changes little as n is varied (accounting for the results in Figure 3.19), whilst the second and third peaks rapidly damp with increasing softness. A successful perturbation theory should, therefore, be able to reproduce the large-k region of the structure factor.

The exclusion

$$\xi(r) = (1 + f(r)) = \exp\left(-\frac{\Phi(r)}{kT}\right)$$

differs in the hard- and soft-sphere cases as shown in Figure 3.30(a), and their difference $\Delta\xi = \xi_s - \xi^0$, where ^{0} always refers to the hard-sphere reference system, is shown in Figure 3.30(b).

Anderson, Weeks and Chandler obtain a functional Taylor expansion in terms of $\Delta\xi$ about a hard-sphere reference system for the free energy of soft spheres:

$$F_s = F^0 + \int \frac{\partial F^0}{\partial \xi(r)} \Delta\xi(r)\, dr + \frac{1}{2!} \iint \frac{\partial^2 F^0}{\partial \xi(r)\partial \xi(r')} \Delta\xi(r)\, \Delta\xi(r')\, dr\, dr' + \ldots \quad (3.97)$$

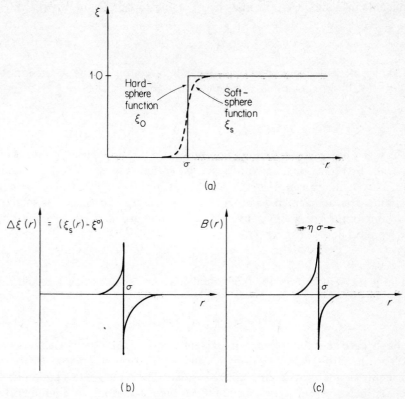

Figure 3.30 (a) The exclusion function $\xi = 1 + f(r)$ for hard and soft spheres. (b) The difference $\Delta\xi = \xi_s - \xi_0$ between the hard and soft functions. (c) The blip function $B(r)$ described in the text

From F_s we can, of course, calculate the pressure, entropy and other thermodynamic functions by straightforward differentiations with respect to ρ and β. The radial distribution function can also be obtained by functional differentiation with respect to $\xi(r)$:

$$\tfrac{1}{2}\rho^2 g_{(2)}(r) = \xi(r)\frac{\partial F}{\partial \xi(r)} \tag{3.98}$$

Anderson *et al.* work in terms of a function $y(r)$ related to $g_{(2)}(r)$ as

$$y(r) = \exp(\beta\Phi(r))g_{(2)}(r)$$

so that (3.98) becomes

$$y(r) = \frac{2}{\rho^2}\frac{\partial F}{\partial \xi(r)}$$

If we now define a 'blip' function $B(r)$ (Figure 3.30c)

$$B(r) = y^0(r)[\xi_s(r) - \xi^0(r)] = y^0(r)\,\Delta\xi(r) \tag{3.99}$$

then from (3.98) and (3.99) the free-energy expansion (3.97) can be written

$$F_s = F^0 + \tfrac{1}{2}\rho^2 \int B(r)\,\mathrm{d}r + \ldots \qquad (3.100)$$

The range of $B(r)$ is very small and depends of course on the softness of the sphere. We choose the atomic diameter σ such that

$$\int_{-\infty}^{\infty} B(r)\,\mathrm{d}r = 0 \qquad (3.101)$$

(σ is in fact a decreasing function of both temperature and density). This choice causes the first functional derivative, and potentially the largest term, to vanish identically. If the range of $B(r)$ is $\eta\sigma$, then it may be shown that the second functional derivative enters to order η^4 for this choice of σ. Since η is small the expansion for the free energy converges rapidly and provides a direct connection between hard and soft spheres:

$$F_s = F^0[1 + O(\eta^4)] \qquad (3.102)$$

Similarly the structure of the 'soft' fluid may be related to that of the rigid sphere:

$$g_s(r) = \xi_s(r)y^0(r)[1 + O(\eta^2)] \qquad (3.103)$$

although here the convergence is evidently less rapid. Analytic expressions exist for the hard-sphere functions F^0 and $y^0(r)$—these summarize the results of machine simulations of hard-sphere systems. The results of this approach are excellent, being within 1 per cent of the Monte Carlo results at high densities, and at all densities at high temperatures when the core of the pair potential dominates the interaction.

References

C. A. Croxton, *LSP*.

G. H. A. Cole, *An Introduction to the Statistical Theory of Classical Simple Dense Fluids*, Pergamon (1967), Chapters 4, 5.

J. G. Kirkwood, *Theory of Liquids*, Gordon and Breach. A collection of Kirkwood's papers on the structural properties of liquids.

I. Z. Fisher, *Statistical Theory of Liquids*, Chicago (1961), Chapters 2–5.

CHAPTER 4

Numerical Solution of the Integral Equations

Introduction

Numerical solution and assessment of the various structural theories of fluids may be conveniently divided into three sections. First of all the *low-density solutions* to the integrodifferential equations will be discussed. Comparison with experiment and the exact results of Chapter 2 is not made in terms of the radial distribution which is relatively structureless at these densities, but instead in terms of the virial coefficients. In as far as the pairwise additivity of the molecular interaction is an acceptable approximation, the virial coefficients determined in Chapter 2 represent the essentially *exact* results, and it is these results which form the basis for comparison with the approximate theories.

At higher fluid densities, characteristic of the condensed liquid phase, comparison is made both in terms of the structural and thermodynamic features of the system. Ultimately comparison must be made with real, simple fluids: unfortunately this does not simply mean the insertion of a realistic pair potential in the integral equations. The discrepancies between the experimental data and theory are quite considerable, and there is some difficulty in deciding whether the inadequacy is to be attributed to the theory, the assumed form of pair potential, the method of numerical solution, the data or a combination of all these sources of uncertainty. For this reason solutions for idealized forms of potential—hard sphere or square well—are compared with molecular dynamics or Monte Carlo data generated in computer simulation. Apart from the computational convenience of these model interactions, machine simulations of these idealized assemblies is straightforward and the ambiguities regarding the nature and/or interpretation of the experimental data, the form of the pair potential and so on do not arise. Direct comparisons and assessments of the adequacy of the structural theories may then be made without the intervening uncertainties. We shall therefore devote some attention to physically unrealistic fluids in one, two and three dimensions in addition to more realistic assessments.

Finally, the various integral equations relating the structure to the pair potential may, of course, be used 'in reverse' to determine the form of the intermolecular potential. Naturally, inversion of the approximate equations engenders some uncertainty in the final potential function; useful information should be obtainable nevertheless, particularly in the case of the liquid metals for which we are theoretically led to anticipate a fundamentally different long-range form to those of insulating fluids.

In the course of approximation internal inconsistencies are inevitably introduced with the consequence that the pressure and compressibility equations of state

(p)
$$\frac{P}{\rho kT} = 1 - \frac{2\pi\rho}{3} \int_0^\infty g_{(2)}(r)\nabla\Phi(r)r^3 \, dr$$

(c)
$$\frac{1}{kT}\left(\frac{\partial P}{\partial \rho}\right)_T = 1 - 4\pi\rho \int_0^\infty c(r)r^2 \, dr$$

(4.1)

tend to bracket the correct isotherm. The self-consistent approximations (SCA) of Rowlinson and Hurst have been discussed (p. 97) and these attempt to *force* thermodynamic consistency by taking density-dependent interpolations between the bracketing isotherms. The remaining theories are inconsistent and we shall designate by the subscripts p and c estimates based on the pressure and compressibility equations of state, respectively.

We begin first of all with a comparison of the low-density results on the basis of the various theories developed in Chapter 3 with the essentially exact results of Chapter 2.

Low-density Solutions: Hard Spheres

One Dimension

The Percus–Yevick equation yields the exact equation of state for an array of rods constrained to move on a line. Since the PY approximation retains the exact short-range form

$$c_{PY} = f(1 + C) \tag{4.2}$$

where f and C represent the Mayer f-function and chain diagrams respectively, it is clear that since there can be no indirect correlation communicated between rods 1 and 2 through any indirect routes, $c(r) = f(r)$ exactly in this case. The equation of state may be shown to be

$$\frac{PV}{NkT} = (1 - \alpha)^{-1}, \qquad \alpha = \frac{\rho b}{N} \tag{4.3}$$

where b/N is the length of one rod. The exact virial coefficients follow immediately:

$$B_2/b = B_3/b^2 = B_4/b^3, \text{etc.} = 1 \tag{4.4}$$

and these results are compared with the estimates based on the HNC approximation in Table 4.1. From the results in Table 4.1 we observe the usual bracketing of the exact result by the inconsistent HNC pressure and compressibility coefficients. From equation (4.1) we can conclude that the HNC approximation overestimates the amplitude of the principal peak in $g_{(2)}(r)$ for hard rods. This will prove a general feature of the HNC structural solutions. From the compressibility isotherm we may conclude that the positive tail to the direct correlation together with the underestimate of the negative core (3.61) results in $(\partial P/\partial \rho)_T$ being too small, and hence $(B_n/b^{n-1})_c$ representing an underestimate

Table 4.1. Virial coefficients for hard spheres in
one dimension (rods)

	B_4/b^3	B_5/b^4
HNC_p	3/2	11/6
HNC_c	11/12	4/5
$PY_{p,c}$(Exact)	1	1

of the exact virial. The positive tail should not develop at all, of course. The tail is attributed to indirect correlations communicated between particles 1 and 2 and these are explicitly forbidden in a one-dimensional model. This is precisely why PY, in dropping the tail and retaining the exact core, is able to generate the exact equation of state.

Two Dimensions

The equation of state is not known for hard discs. The second and third virial coefficients are known exactly whilst the fourth, fifth and sixth have been determined by Ree–Hoover evaluation of the cluster integrals (Chapter 2).

The cluster integral representation of the virial coefficients physically describes the contribution of two-body, three-body, ... configurations of particles to the equation of state, and in this way accounts for the departure from idealism of the system. We see from equations (3.64) and (3.65) that in the restricted PY and HNC diagrammatic sums approximation first enters at order ρ^2. This means that the two- and three-body contributions are retained complete, and therefore in two and three dimensions we should expect both the HNC and PY approximations to yield B_2 and B_3 exactly. The results are shown in Table 4.2.

Table 4.2. Virial coefficients for hard spheres in two dimensions (discs) $(b = \pi\sigma^2/2)$

	B_2/b	B_3/b^2	B_4/b^3	B_5/b^4	B_6/b^5
Exact	1	0·7820	0·5322	0·3338	0·1992
HNC_p	1	0·7820	0·8066		
HNC_c	1	0·7820	0·4423		
PY_p	1	0·7820	0·5008		
PY_c	1	0·7820	0·5377		

Inconsistency appears at the fourth virial coefficient which represents the first diagrammatic approximation. Bracketing of the true isotherm by the pressure and compressibility curves occurs as usual, although we should notice that the HNC and PY discrepancies are in opposite directions. Thus, either an interpolation between the PY and HNC isotherms (Rowlinson, (3.92)) or between the two HNC (Hurst, (3.93)) or PY curves might enforce self-consistency, and these of course represent the self-consistent approximations.

Apart from the internal inconsistency, the superiority of the PY approximation for hard-disc assemblies at low densities is clearly shown. The core of the direct correlation is correctly represented in the PY approximation: the discrepancy may be attributed to the neglect of indirect correlation which may develop between particles 1 and 2 through adjacent discs. The neglect of indirect correlations does not arise in one dimension: as we might anticipate, the discrepancy will be worse in three dimensions.

HNC misrepresents both the core and the long-range tail of the direct correlation, and although the latter should undoubtedly develop for all dimensions > 1, the underestimate of the core engenders a serious discrepancy at all densities.

Three Dimensions

Not surprisingly, the main effort has been directed at the determination of the hard-sphere virial coefficients, and as we see from Table 4.3, these are inconsistent beyond B_3.

Table 4.3. Virial coefficients for hard spheres in three dimensions

	B_2/b	B_3/b^2	B_4/b^3	B_5/b^4	B_6/b^5
Exact	1	0·625	0·2869	0·1103	0·0386
BGY_p	1	0·625	0·2252	0·0475	
BGY_c	1	0·625	0·3424	0·1335	
K_p	1	0·625	0·1400		
K_c	1	0·625	0·4418		
$HNC1_p$	1	0·625	0·4453	0·1447	0·0382
$HNC2_p$	1	0·625		0·066	
$HNC1_c$	1	0·625	0·2092	0·0493	0·0281
$HNC2_c$	1	0·625		0·123	
$PY1_p$	1	0·625	0·2500	0·0859	0·0273
$PY2_p$	1	0·625		0·124	
$PY1_c$	1	0·625	0·2969	0·1211	0·0449
$PY2_c$	1	0·625		0·107	

The analytic pressure and compressibility hard-sphere equations of state in the Percus–Yevick approximation are

$$\left(\frac{PV}{NkT}\right)_p = \frac{1 + 2\eta + 3\eta^2}{(1 - \eta)^2}$$
$$\left(\frac{PV}{NkT}\right)_c = \frac{1 + \eta + \eta^2}{(1 - \eta)^3} \qquad \eta = \tfrac{1}{6}\pi\sigma^3\rho \tag{4.5}$$

from which it is straightforward to show that

$$(B_l)_p = 2(3l - 4)(b/4)^{l-1}$$
$$(B_l)_c = [1 + \tfrac{3}{2}l(l - 1)](b/4)^{l-1} \tag{4.6}$$

for $l \geqslant 2$. $(B_l)_p$ and $(B_l)_c$ are seen to be identical for $l = 2, 3$, and to differ slightly thereafter (Table 4.3, Figure 4.1). The exact hard-sphere coefficients B_4, B_5 and B_6 have been evaluated from the Ree–Hoover cluster integrals in Chapter 2.

Although it has only recently been shown (Croxton, 1975), it is nevertheless evident from Table 4.3 that the pressure and compressibility curves bracket the correct isotherm (Figure 4.1). PY retains its superiority in both the first- and second-order theories; the neglect of the long-range tail of the direct correlation function in the highly connected fluid accounts however for the inferiority of the three-dimensional results in comparison to the two-dimensional coefficients in which the tail plays a subordinate role, and the one-dimensional system in which is it non-existent. Nevertheless, the cancellation scheme in the Rush-brooke–Scoins expansion of the direct correlation ensures that PY gives virtually the exact function for hard spheres, despite the apparently more drastic nature of this approximation.

The ratios B_l/B_l (exact) for the various approximations in Table 4.3 are summarized in Figure 4.1(b) from which the general divergence of the higher-order virial coefficients is clearly apparent for $l > 3$, as is the bracketing by the compressibility and pressure estimates. The curious behaviour of HNC_p and HNC_c is presumably to be attributed to particularly felicitious cancellation amongst the higher-order diagrams.

The self-consistent approximation of Rowlinson, equation (3.92), remains both exact and consistent up to B_3, but thereafter remains only self-consistent, Table 4.4. Hurst's GHNC approximation, however, remains virtually exact up

Table 4.4

	B_4/b^3	B_5/b^4	B_5/b^5
Exact	0·2869	0·1103	0·0386
Rowlinson SCA	0·2824	0·1041	0·0341
Hurst GHNC	0·2824	0·1102 (c)	0·0386 (c)
		0·0915 (p)	0·0353 (p)

to B_6, but develops minor inconsistencies at and beyond B_5. Hurst used a value of $m = 0.4372$ in equation (3.93): Henderson has subsequently shown that a minor modification in m ensures the *simultaneous* consistency of B_4, B_5 and B_6. These self-consistent theories provide little physical insight into the various approximations, although some diagrammatic justification for the SCA of Rowlinson can be given. Otherwise, these approximations represent a simple interpolation of the useful features of the HNC and PY approximations (and of their shortcomings): their justification lies primarily in their usefulness as accurate, self-consistent low-density equations of state. They do show, how-ever, that an exact theory must presumably exhibit the intermediate charac-teristics of the HNC and PY direct correlations simulated by the Hurst and Rowlinson interpolations.

108

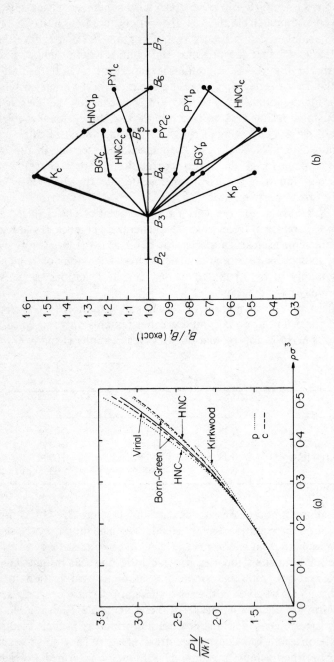

Figure 4.1 (a) The low-density hard-sphere equation of state in the various approximations. (b) The ratio B_l/B_{exact} of the virial coefficients in the various approximations. All theories are seen to be consistent up to B_3, and then the inconsistent pressure and compressibility estimates tend to diverge and bracket the exact value

Low-density Solutions: Lennard–Jones Systems

The virial coefficients for more realistic potentials such as the square-well and Lennard–Jones interactions are of course temperature dependent. Different regions of the pair potential dominate the interaction at different temperatures and in consequence there are relatively few results for these more extensive computations. The diagrammatic completeness of the first-order HNC and PY approximations ensures that $B_2(T)$ and $B_3(T)$ are both exact and self-consistent. Beyond that, diagrammatic approximation engenders inconsistency between the pressure and compressibility estimates.

At high temperatures, when the core of the pair potential dominates, the PY approximation yields superior results as we might expect. At lower temperatures none of the approximations is satisfactory although PY generally gives the better representation. In Figure 4.2 we compare the reduced fourth

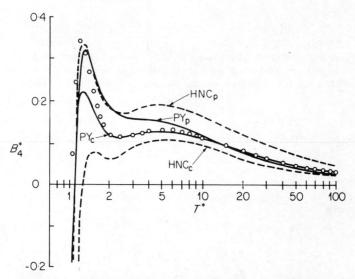

Figure 4.2 The reduced fourth virial coefficient at high temperatures. The circles give the exact values and the curves give the results of the PY and HNC theories in the pressure (p) and compressibility (c) approximations

virial coefficient for a Lennard–Jones system on the basis of the HNC and PY theories with the exact results of Barker (see also Figure 2.8). The results bracket the correct virial as usual, although HNC remains significantly discrepant at high temperatures.

The fifth virial coefficient determined on the basis of the PY, PY2, HNC, HNC2 and SCA approximations are shown in Figures 4.3 and 4.4. HNC_c is seen to be particularly bad, and is not even qualitatively correct since it has only one maximum. Both $HNC2_c$ and $PY2_c$ are seen to be good over the entire

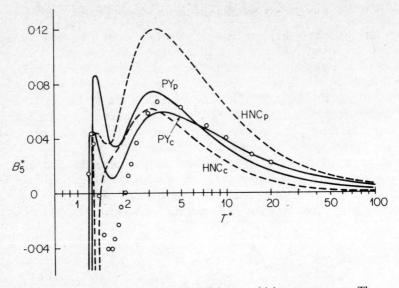

Figure 4.3 Reduced fifth virial coefficient at high temperatures. The circles give the exact results of Barker, and the curves give the results of the HNC and PY theories in the pressure (p) and compressibility (c) approximations

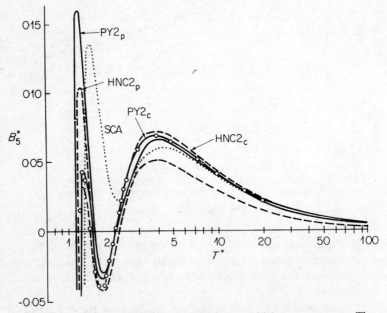

Figure 4.4 Reduced fifth virial coefficient at high temperatures. The circles give the exact results of Barker, and the curves give the results of the HNC2 and PY2 theories in the pressure (p) and compressibility (c) approximations. Also shown is the self-consistent approximation (SCA)

temperature range of B_5 although the second-order pressure coefficients completely misrepresent the low-temperature maximum. Nevertheless, all the second-order theories obtain the pronounced minimum in $B_5(T)$ in the region $1.4 < T^* < 2$, which none of the first-order theories manage. The SCA estimate of Hurst and of Rowlinson for $B_5(T)$ shown in Figure 4.4 are graphically indistinguishable: the results are seen to be disappointing.

In Table 2.4 we quoted the critical ratios for a Lennard–Jones fluid estimated on the basis of the *exact* five-term virial. We should point out that these do not represent the exact critical ratios however. Moreover, Barker obtained his critical ratios on the basis of the first and second derivatives of the equation of state. The critical point is highly singular, and the taking of derivatives must be approached with some caution. Nevertheless, in Table 4.5 we compare these

Table 4.5 Critical ratios for a Lennard–Jones fluid ($b = 2\pi\sigma^3/3$)

	kT_c/ε	b/V_c	P_cV_c/kT_c
'Exact'	1.291	0.547	0.354
HNC_p	1.25	0.98	0.35
HNC_c	1.39	0.91	0.38
PY_p	1.25	0.88	0.30
PY_c	1.32	0.91	0.36
$PY2_p$	1.36	0.73	0.31
$PY2_c$	1.33	0.77	0.34

results with the HNC, PY and PY2 critical ratios. The experimentally observed critical ratio kT_c/ε is 1.26. Curiously PY2 shows a *worsening* of agreement, and the critical volume is particularly bad for all theories.

Liquid-density Solutions

Some caution must be exercised in anticipating the performance of the various structural theories on the basis of their low-density performance. Term-by-term comparison of virial coefficients effectively represents order-by-order comparison of the diagrams retained in the various approximations to the Rushbrooke–Scoins expansion of the direct correlation. At liquid densities, however, we work simultaneously with diagrams to *all* orders of density and the extensive cancellation which develops amongst the diagrams cannot be anticipated from the virial coefficients. In the case of the Kirkwood superposition approximation, for example, we anticipated that this closure device should have limiting validity only as the density tended to zero. *A priori* considerations suggested that such an approximation to triplet correlation in a dense fluid would be quite inappropriate. In fact, both on the basis of machine simulations and Salpeter expansion of $g_{(3)}(123)$, we found that the superposition approximation was surprisingly satisfactory, and this could be directly attributed to extensive diagrammatic cancellation in the Salpeter series, leaving the super-position product as the leading term in the expansion.

Comparison of the structural and thermodynamic features of dense fluid assemblies will generally be made with computer simulations: this eliminates any ambiguity arising from uncertainties in the pair potential. Obviously such comparisons are essential in the case of idealized assemblies of particles inter-acting through square-well and hard-sphere potentials. The machine simulation methods will be discussed fully in Chapter 10; however, it is appropriate to outline the two principal methods.

It is possible to solve simultaneously the Newtonian equations of motion for a limited number of interacting particles $\sim 10^3$ for a three-dimensional system, considerably more in two dimensions. The structure and dynamics of the system then develop as time averages over the evolving distributions: this is the *molecular dynamics* method. Alternatively the *Monte Carlo method* is used in which a Markov chain of successive states is generated in such a way that a given state will occur with a frequency proportional to its probability in the ensemble as the chain length is increased. Thus each state receives its correct weighting and the properties of interest arise as averages over configuration space: dynamical information cannot be obtained in this method, however.

Hard-sphere Solutions

Insertion of the hard-sphere potential in the pressure equation

$$\frac{PV}{NkT} = 1 - \frac{2\pi N}{3VkT} \int_0^\infty \nabla\Phi(r)g_{(2)}(r)r^3 \, dr \tag{4.7}$$

yields the particularly simple equation of state

$$\frac{PV}{NkT} = 1 + \frac{2\pi\sigma^3}{3}\rho g_{(2)}(\sigma) \tag{4.8}$$

where $g_{(2)}(\sigma)$ is the height of the first peak of the pair distribution at $r = \sigma$. $g_{(2)}(\sigma)$ is of course a function of density itself, but because of the singular nature of the potential core, is not a function of temperature. The numerical solution of the various integral equations for the pair correlation is a task of considerable magnitude, and we shall not go into the details here.

The KBGY class of equations in the superposition approximation were the first to be numerically evaluated for hard spheres. The radial distribution function shows the correct qualitative features with increasing density: the amplitude and sharpness increasing with increasing density (Figure 4.5). $g_{(2)}(\sigma)$ determined in this way may be substituted in equation (4.8) to yield the equation of state (Figure 4.6). The BGY and K equations are by far the worst numerically, but have the qualitative advantage that this class fails to yield solutions beyond some critical density. In the case of the BGY the critical density is $\rho\sigma^3 = 0.95$, whilst for the Kirkwood equation solutions cannot be obtained beyond $\rho\sigma^3 = 1.14$. This critical density was speculatively identified with the limit of

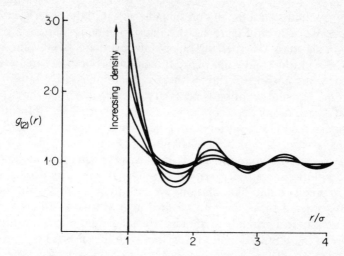

Figure 4.5 The hard-sphere radial distribution function in the BGYK approximation as a function of density. The curves show the expected qualitative features: a region of particle exclusion within the collision diameter σ, and radially damped oscillations whose amplitude decreases with decreasing density

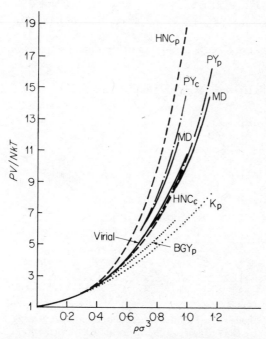

Figure 4.6 The equation of state of the hard-sphere fluid in the various approximations

stability of the fluid solutions, and presumably signalled the onset of crystalline solutions. As we see from Figure 4.6, this interpretation is substantiated by the machine simulations. Whether this breakdown in the KBGY solutions may be identified with the solid–fluid phase transition has been debated ever since with no rigorous answer. Certainly there is some qualitative appeal in this identification, but some caution is necessary: the singularity in the *two-dimensional* KBGY equations does not coincide with the simulated transition; and more seriously, KBGY continues to predict a phase transition in a one-dimensional system for which there is known to be no such phase change.

In spite of these reservations, the KBGY class of equations alone cease to yield fluid solutions beyond a critical density—all other first-order theories continue to predict liquid-like solutions even at physically impossible densities. The second-order theories have not yielded high-density solutions as yet, but there are no physical grounds to anticipate a solid–fluid phase transition. It is this feature of the KBGY equations together with the surprisingly satisfactory nature of the Kirkwood superposition approximation which suggested its refinement in terms of the Salpeter expansion, equation (3.34). The KBGY equations are, of course, exact prior to the Kirkwood closure and Rice and Lekner have computed the $P(3, 3)$ Padé approximant to the exact triplet correlation $g_{(3)}(123)$ (equation (3.37)). Their results for hard spheres are shown in Figure 4.7. The Padé approximant to the triplet correlation

$$g_{(3)}(123) \sim g_{(2)}(12)g_{(2)}(23)g_{(2)}(31) \cdot \frac{\rho\delta_4(123)}{1 - \rho\delta_5(123)/\delta_4(123)} \qquad (4.9)$$

is seen to yield a much better result than the linear expansion of the Salpeter series (Figure 4.7)

$$g_{(3)}(123) \sim g_{(2)}(12)g_{(2)}(23)g_{(2)}(31)[\rho\delta_4(123) + \rho^2\delta_5(123)] \qquad (4.10)$$

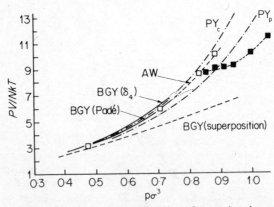

Figure 4.7 Hard-sphere equation of state, showing the BGY (Padé) isotherm to be in essential agreement with the machine simulations (□ 108 particles, fluid; ■ 108 particles, solid)

This is not surprising since the cluster integrals δ_4, δ_5,... alternate in sign and the cancellation is better simulated in (4.9) rather than (4.10) where δ_4 dominates. Higher-order Padé approximants would presumably yield even closer agreement with the machine simulations: the computational labour involved in evaluating the higher coefficients δ_6,... is, however, prohibitive.

The results of the BGY–Padé approximation are compared with the five- and six-term virials and with the molecular dynamics results in Table 4.6. This

Table 4.6

$\rho\sigma^3$	Z_5	Z_6	BGY–Padé	Molecular dynamics
0·88	8·11	8·95	10·11	10·17[a]
0·83	7·17	7·79	8·55	8·59
0·71	5·31	5·59	5·83	5·89
0·47	2·98	3·01	3·03	3·05
0·14	1·36	1·36	1·36	1·36

[a] Density transition occurs at $\rho\sigma^3 = 0.87$.

approach has been applied to Lennard–Jones fluids with some success, although the Padé–Salpeter representation of the triplet correlation is less satisfactory for the reasons discussed in Chapter 3 (p. 67). We should note that no solutions are obtained beyond the critical density. The breakdown in the KBGY equations presumably cannot therefore be attributed to the inadequacy of the superposition approximation as is sometimes suggested.

The analytic equations of state for a hard-sphere fluid in the Percus–Yevick approximation given in equation (4.5) are shown in Figure 4.6 and are seen to be in good agreement with the fluid branch of the machine isotherm. There is, however, no indication of a phase transition although the equations of state become singular at $\eta = 1$, i.e. $\rho\sigma^3 \sim 1.9$ which is impossibly greater than the close-packed density of 1·41. Carnahan and Starling have proposed the empirical hard-sphere relation

$$\frac{PV}{NkT} = \frac{1 + \eta + \eta^2 - \eta^3}{(1 - \eta)^3} \qquad (4.11)$$

which agrees with the fluid branch of the machine simulations to within graphical accuracy. The Carnahan–Starling isotherm is in striking agreement with the analytic PY hard-sphere isotherms, equations (3.73), (3.74) (Table 4.7).

For strongly repulsive potentials the PY approximation appears to give the best results over the fluid density range. We have seen that PY retains the exact core of the direct correlation, but neglects the long-range tail (Figure 4.8a). For the hard-sphere interaction this amounts to setting

$$B' + E = 0$$

It is quite straightforward to show that for hard spheres $B' + E \sim 0$ and to this extent the PY approximation is vindicated. For example, the first B' and

116

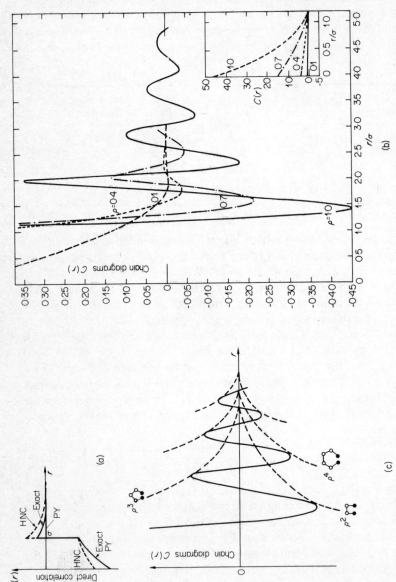

Figure 4.8 (a) Comparison of the PY, HNC and exact hard-sphere direct correlations. The PY approximation is exact within the collision diameter, but neglects the long-range tail. HNC, whilst developing a long-range tail is inexact over both regions. (b) The chain diagrams $C(r)$ as a function of density. Clearly these diagrams are responsible for the oscillatory form of the total correlation and RDF. (c) Schematic contribution of progressively higher-order chain diagrams on the sustained oscillatory structure of $C(r)$.

Table 4.7

η	PV/NkT Carnahan–Starling	PY_p	PY_c
0·1	1·52	1·52	1·52
0·2	2·40	2·37	2·42
0·3	3·97	3·82	4·05
0·4	8·77	6·33	7·22
0·5	13·00	11·00	14·00
0·6	27·25	20·50	30·63
0·7	68·40	43·00	81·11

E diagrams dropped are ⬡ and ⬡ : since the number of f-bonds is even in the first diagram and odd in the second it is clear that they are of opposite sign since $f(ij) = -1\cdot0$ over its effective range. Moreover, from Figure 3.18 we see that they are of almost identical magnitude and cancel almost completely. Of course, it does not follow that *all* the higher-order B' and E diagrams will cancel in this way, indeed they do not. But it may be shown (see Chapter 5) that a significant proportion do cancel and the long-range tail of the hard-sphere direct correlation is weak.

From equation (3.67) we saw that

$$g_{(2)}(12) - c(12) = 1 + C$$

and since for hard spheres $c(12) = 0$ for $r > 0$ in the PY approximation, the long-range form ($r > \sigma$) of the pair distribution function is evidently given as $(1 + C)$. The sum over chain diagrams $C(12)$ for hard spheres is shown in Figure 4.8(b) for a range of densities. Quite clearly the periodic behaviour of the radial distribution function comes from these diagrams, and even for the HNC approximation in which $c(12)$ exhibits a weak positive long-range tail, the structure will be dominated by the oscillations in $C(12)$, although they will be enhanced by $c(12)$ relative to the PY radial distribution. The chain diagrams, having no direct Mayer $f(12)$ bond by definition, evidently represent the communication of correlation between particles 1 and 2 through chains of neighbouring particles. Obviously the *range* of a chain diagram will depend upon the number of f-bonds: ●–○–○–●, ●–○–○–○–●, etc., whilst the *sign* ($\pm$) of the diagram depends on whether the number of bonds is odd ($C(12) +$ve) or even ($C(12) -$ve). At a given radial separation a given order of chain diagram will dominate: this is shown schematically in Figure 4.8(c).

The potential of mean force has been given in equation (3.55) which, in the HNC approximation ($E(12) = 0$) reduces to

$$\frac{\Psi(12)}{kT} = \frac{\Phi(12)}{kT} - C(12)$$

Since $\Phi(12) = 0$ for $r > \sigma$ for hard spheres there is no direct force between two

118

particles when they are separated by more than a hard-sphere diameter. Since $C(12) = 0$ however, there is an indirect (or statistical) force resulting from inter-actions communicated through other particles. The large values of $C(12)$ shown in Figure 4.8(c) for $r < \sigma$ do not affect the potential of mean force since $\Phi(12) = \infty$ for $r < \sigma$ and hence $\Psi(12) = \infty$ independent of $C(12)$.

Since the Percus–Yevick approximation describes the fluid branch as well as it does, it is in some ways surprising that there is no indication whatsoever of a phase transition: the equation of state is a monotonic increasing function of density and becomes singular only far beyond the physical maximum packing density. Temperley has shown that solutions to the PY equation of the form $\cos(\alpha_i r)$ exist, where the α_i have the significance of reciprocal distances charac-teristic of the structure of the system. Hutchinson has shown that these solutions unfortunately imply negative intensities of radiation scattering at certain angles, and that these results cannot therefore be accepted. The relation between the structure factor $S(k)$ and the Fourier transform of the direct correlation $c(k)$ is (equation (3.78))

$$S(k) = \frac{1}{1 - \rho c(k)} \tag{4.12}$$

At high densities the oscillations in $c(k)$ become very pronounced and account for the near-singular form of $S(k)$ for periodic structures (Figure 4.9), i.e. $S(k)$ adopts the form of an array of δ-functions. It is clear that at these densities

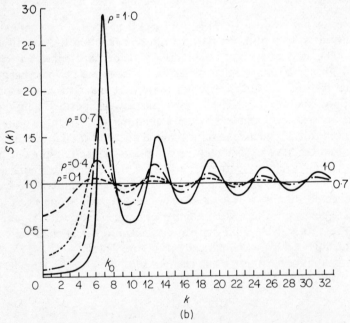

(b)

Figure 4.9 Development of the hard-sphere structure factor with density in the HNC approximation

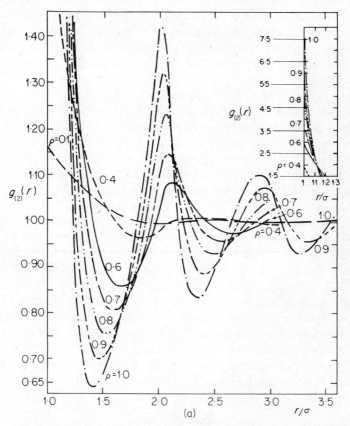

Figure 4.10 Development of the radial distribution function with density in the HNC approximation

there is little room for approximation in $c(k)$, and $S(k)$ is correspondingly susceptible to inaccuracies in the direct correlation. If for any reason $\rho c(k) > 1$, then $S(k)$ adopts negative values and may be dismissed as being physically unacceptable. This is a harsh criterion and probably some relaxation in its application will be necessary before a theory of phase transitions can emerge. The phase transition aspect of the PY equation will be discussed in more detail in Chapter 5.

The HNC approximation again yields a radial distribution function of the correct qualitative form: the oscillations in $g_{(2)}(r)$ develop in both amplitude and sharpness as the density increases. Similarly the principal peak in $S(k)$ shows a pronounced development with density indicative of long-range structuring. The subsidiary peaks imply an increase in the fine structure of $g_{(2)}(r)$, Figure 4.10.

Of particular interest in these $S(k)$ curves, not only in the HNC case, is the long-wavelength form of the structure factor, i.e. $S(k)$ as $k \to 0$. As the oscillations diminish in amplitude with decreasing density, it is a consequence of the sum rule, equation (3.79), that $S(0)$ *increases*, and from the thermodynamic relation $S(0) = \rho k T \chi_T$ it follows that the compressibility also increases. In particular, singular behaviour at $k = 0$, $\rho > 0$ implies infinite compressibility or, what is the same thing, zero slope of the rigid-sphere isotherm at that density, and this kind of collective behaviour is what would be required at a first-order phase transition. No such tendency towards a long-wavelength singularity is observed in the HNC approximation, and there is presumably no chance of a phase transition.

Diagrammatically we may understand the inadequacy of the HNC approximation in terms of the classes of graph summed. Both the short-range and the long-range form of $c_{HNC}(r)$ are inexact for hard spheres, equation (3.61) and Figure 4.8. Certainly, the compressibility equation of state (equation (3.87)) is relatively insensitive to inaccuracies in the form of the core of the direct correlation. It is, however, highly sensitive to the long-range tail, and its overestimate in HNC and its neglect in PY accounts for the departure of both these compressibility curves from the simulated isotherm (Figure 4.6).

Realistic Fluids

In the case of realistic fluids, comparison between theory and experiment is obscured by our inadequate knowledge of the pair potential operating within the fluid. Certainly three-body effects intervene and the pair potential is weakly density dependent even for the simple liquids. Nevertheless, an 'effective' pair potential may be used and this is generally different to the theoretically computed pair potential operating between an isolated pair of classical atoms. For this reason comparison is often made with machine simulations of simple classical fluids in which case there can be no ambiguity in the pair potential.

In Figure 4.11 we show the Monte Carlo and experimental $T^* = kT/\varepsilon = 2.74$ isotherm for liquid argon: the curves are in very good agreement and either

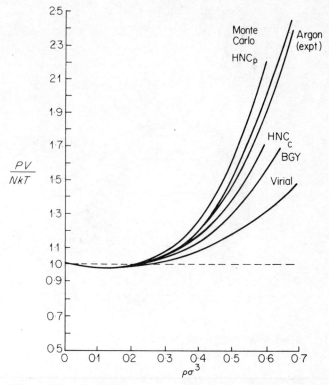

Figure 4.11 Comparison of the liquid argon equation of state in the various approximations

may be used as the reference isotherm. The Lennard–Jones potential with $\varepsilon/k = 119.8$ K and $\sigma = 3.405$ Å is used in the simulation and theoretical calculations. Broyles, Chung and Sahlin have solved the PY, HNC and BGY equations on the $T^* = 2.74$ isotherm and find that PY agrees most satisfactorily with the equation of state and its prediction of the internal energy. HNC is next best and BGY the worst.

For realistic fluids the long-wavelength limit ($k \to 0$) of the structure factor as defined in equation (4.12) will show singularities for certain combinations of density and temperature values, defining the limit of stability in the ρT plane. Such singularities imply $(\partial P/\partial \rho)_T = 0$. Unfortunately, for all points on the experimental coexistence curve, $(\partial P/\partial \rho)_T \neq 0$ except at the critical point, and for this reason the theoretical locus of singularities *cannot* be identified with the experimental coexistence curve. The shaded region in Figure 4.12 corresponds to physically accessible supercooled or superheated states of the fluid which are relatively less stable than the competing states at the same temperature and pressure: the region within the broken locus represents physically unattainable states corresponding to mechanically unstable situations. The Klein–Green locus corresponds to the limit of mechanical stability of the system.

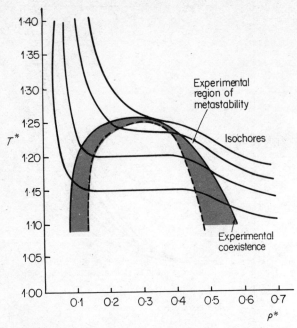

Figure 4.12 Region of liquid–vapour coexistence in argon. The broken curve represents the theoretical locus of $(\partial P/\partial \rho)_T = 0$ in the HNC approximation. The full curve represents the experimental coexistence boundary.

Inversion of the Integral Equations

The various integral equations relating the pair distribution to the pair potential may, of course, be inverted. Thus, inserting the experimental data for $g_{(2)}(r)$ we may determine $\Phi(r)$. There are reasons to believe that the liquid-metal pair potentials are fundamentally different from their liquid-inert-gas counterparts. In particular, Friedel screening of the ionic core by conduction electrons in a liquid metal implies a long-range oscillatory (LRO) pair interaction whilst the dispersion forces between inert-gas atoms result in a monotonic attractive branch of range ~ two or three atomic diameters. The LRO interaction was first anticipated theoretically and inversion of the integral equations provided the first evidence of oscillations.

We have seen that $c(r)$ has, by definition, approximately the range of the pair potential. It is evident from the localized form of the quantity $c(k)/c(0)$ in the small-k region that the liquid-metal interactions are evidently of much greater range than those of the inert gases (Figure 4.13).

If we rearrange the HNC and PY approximations, equations (3.60) and (3.63), we find immediately

$$\frac{\Phi_{PY}(r)}{kT} = \ln\left[1 + h(r) - c(r)\right] - \ln\left[1 + h(r)\right] \tag{4.13}$$

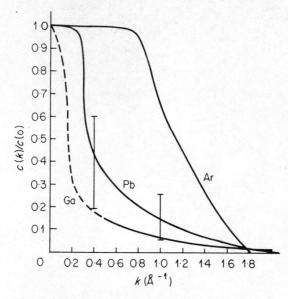

Figure 4.13 Long-wavelength direct correlation ratio $c(k)/c(0)$. The liquid-metal ratios are much more localized than the inert gas function indicating a long-range real space direct correlation

$$\frac{\Phi_{\mathrm{HNC}}(r)}{kT} = h(r) - c(r) - \ln\left[1 + h(r)\right] \tag{4.14}$$

both of which yield the asymptotic form $(h(r) \sim 0)$

$$c(r) \rightarrow -\Phi(r)/kT \tag{4.15}$$

At the zeros of $h(r)$ we have

$$c_{\mathrm{PY}}(r) = -\exp\left(\Phi(r)/kT\right) \tag{4.16}$$

$$c_{\mathrm{HNC}}(r) = -\Phi(r)/kT \tag{4.17}$$

We should, however, be cautious in applying these asymptotic relationships. For example, the *exact* direct correlation for hard spheres shows a weak positive tail and this is clearly inconsistent with equations (4.13), (4.14) and (4.15).

In Figure 4.14 we compare the *actual* pair potential used in a molecular dynamics investigation with the results obtained through the inversion of the structural equations (4.13) and (4.14). In neither case is the initial potential recovered, and the PY inversion is seen to be particularly bad. From.equations (4.13) and (4.14) the general inequality

$$\Phi_{\mathrm{HNC}} \geqslant \Phi_{\mathrm{PY}}$$

follows, since $x \geqslant \ln(1 + x)$, and this is borne out by the results shown in Figure 4.14.

124

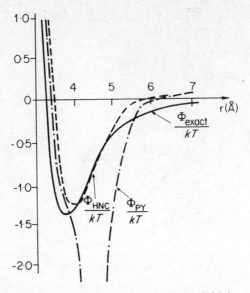

Figure 4.14 Inexact recovery of the initial pair potential $\Phi(r)/kT$ in the PY and HNC approximations

In the case of the controversial liquid-metal pair potentials the question arises as to what extent errors in the scattering data generate artifacts in the inverted potential function. Ballentine and Jones, using the HNC and PY theories, have recently investigated the sensitivity of $\Phi(r)$ to errors in the measured structure factor $S(k)$. Their conclusions for sodium taken as a representative system are that the small-k region of $S(k)$ is by far the most sensitive feature of the structure factor with regard to the determination of pair potentials; the radius of the repulsive core is not sensitive to errors; the depth of the attractive well is the parameter most sensitive to errors in $S(k)$; and the existence of a second repulsive region beyond the first minimum is confirmed.

References

C. A. Croxton, *LSP*.

G. H. A. Cole, *An Introduction to the Statistical Theory of Classical Simple Dense Fluids*, Pergamon (1967), Chapter 6.

CHAPTER 5

Phase Transitions

Introduction

An analytic theory of phase transitions and the location of the boundaries of stability of the liquid phase poses one of the most difficult problems in the statistical theory of fluids and indeed represents one of the major outstanding problems of statistical mechanics. Thermodynamically there is no particular difficulty in the discussion of first- and second-order phase transitions, and Yang and Lee have provided a formal statistical theory of the phase transition. What has not been achieved as yet is an entirely satisfactory physical analysis of phase transitions *at a molecular level*: a theory which predicts not only the macroscopic thermodynamic features of the system, but also yields the microscopic structure of the various phases.

The KBGY class of integral equations, whilst thermodynamically the least satisfactory of the modern statistical theories of fluid structure, was thought to possess the qualitative advantage of signalling a phase transition at some critical density beyond which fluid-like solutions could no longer be obtained. This interpretation was sustained in the case of hard spheres by molecular dynamics simulations in as far as a phase transition was observed at precisely the density the KBGY equations ceased to be integrable. Similar conclusions held for the Lennard–Jones fluids. Unfortunately, as we shall see shortly, the KBGY equations go on to predict a phase transition in two dimensions at a density which is at variance with the machine simulations, and worse, KBGY actually predicts a transition in one dimension where it is known for sure that such a transition cannot occur.

All the other first-order theories continue to predict a homogeneous single fluid phase even beyond physically possible packing densities, and whilst the second-order theories presumably represent an improvement on their first-order counterparts, they remain untested at these densities. There is, however, no physical reason to anticipate any transition phenomena.

If we are to achieve a physical understanding of phase transitions at a molecular level then the physical processes involved must be clearly understood. To this extent formal theories of phase transitions, whilst providing essential criteria in terms of which any adequate theory of phase transitions must be judged, nevertheless arrive at phase transformations for *mathematical* rather than physical reasons. We shall, as far as possible, try to emphasize the physical aspect of the phase transition although we shall inevitably become involved in the formal aspects of this incompletely developed topic.

Equilibrium Between Phases

The PVT surface for a single-component system is shown in Figure 5.1. The boundaries of stability of the various phases are most conveniently shown in

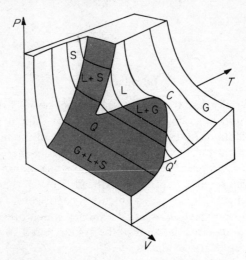

Figure 5.1 PVT surface of a realistic single-component system. The coexistence regions are shown shaded, where G, L and S refer to gas, liquid and solid phases respectively. C and Q are the critical and triple points

the pressure–temperature projection of the surface (Figure 5.2) in which the various domains of relative stability and their boundaries are shown. The PVT surface and its pressure–temperature projection are too familiar to be described in detail here, except to observe that the liquid–gas coexistence boundary QC

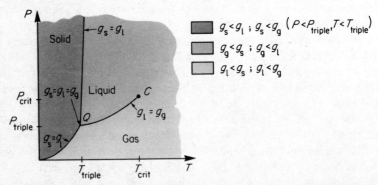

Figure 5.2 The P–T diagram for a realistic single-component system. The liquid–gas coexistence boundary terminates at the critical point C. The solid–liquid coexistence extends indefinitely with no possibility of a continuous transition of the liquid–gas type beyond the critical point

terminates at a critical point beyond which there can be no formal distinction between the gas and liquid phases. This is *not* true for the solid–fluid phase transition. That is, there is no evidence either theoretically or experimentally to suggest that it is possible to pass *continuously* between the solid and fluid phases, implying as it would a solid–fluid critical point.

In the general discussion of the relative stability of two or more phases we recall that the system will adopt that state which *locally* minimizes the Gibbs free energy. It therefore follows for reversible phase transitions that along the coexistence boundaries the molar Gibbs free energy of the adjacent phases must be locally identical. They may therefore coexist in any proportion as we may see from the entire PVT surface. If we designate the two phases i and j, we may consider any point A which lies on the coexistence curve (Figure 5.2) and obtain

$$g_i(P_A, T_A) = g_j(P_A, T_A) \tag{5.1}$$

as the necessary condition for coexistence, where $g_i(P, T)$ represents the molar Gibbs free energy (or the chemical potential) of phase i.

Directly related to the PT projection is the *Clausius–Clapeyron equation* which gives the slope of the phase equilibrium line at any point as

$$\boxed{\frac{\mathrm{d}P}{\mathrm{d}T} = \frac{\Delta S}{\Delta V}} \tag{5.2}$$

where $\Delta S = S_j - S_i$ and $\Delta V = V_j - V_i$ represent respectively the molar entropy and volume changes in the phase transition $j \to i$. Phase transitions $i \leftrightarrow j$ characterized by the discontinuities ΔS and ΔV and the continuity in g at constant pressure and temperature are designated *first-order phase transitions* and are shown schematically in Figure 5.3. We shall be primarily concerned with first-order transitions in this chapter, and we should be able to distinguish these from second-order phase transitions. For these latter transitions $\Delta S = 0$ and $\Delta V = 0$, and consequently are not governed by the Clausius–Clapeyron equation and do not exhibit a horizontal portion on the PV diagram.

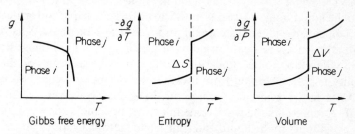

Figure 5.3 Variation of the Gibbs free energy, entropy and volume with temperature across a first-order phase transition. Such a transition is characterized by continuity in the free energy, and discontinuity in the entropy and volume

In most cases of first-order phase transitions an increase in entropy accompanies an increase in volume, thereby asserting the generally positive slope to the coexistence curves. One exceptional case is provided by He^3 in which the nuclear spins are randomly oriented in the solid, whilst over a limited temperature range adopt antiparallel orientations in the liquid phase. Water provides another exception in that the open crystalline structure of ice collapses on melting resulting in a volume *decrease* across the solid–fluid phase transition. In both these cases the solid–fluid coexistence boundary has a negative slope.

The approximate interrelation between the boundaries of stability and the functional form of the pair potential may be quite easily established.

The vaporization curve QC defining the liquid–gas coexistence above the triple point terminates at the critical point C (Figure 5.4a). The critical temperature T_c is roughly proportional to the well-depth ε of the pair potential and

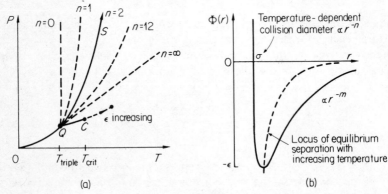

(a)   (b)

Figure 5.4 (a) Dependence of the boundaries of stability on the interaction parameters ε and n describing the depth of the potential well and the repulsive core parameter r^{-n}. (b) Typical pair potential showing the dependence upon the parameters ε and n

in consequence the *range* of liquid–gas coexistence is governed primarily by the magnitude of ε. Thus argon has a liquid range $T_{crit}-T_{triple}$ of only 3 K ($\varepsilon/k = 119$ K) whilst sodium has a liquid range of 782 K ($\varepsilon/k = 600$ K). As $\varepsilon \to 0$ so the liquid range decreases until when $\varepsilon = 0$ the triple and critical points become coincident and the system exhibits no bound liquid phase. A hard-sphere assembly of particles is a case in point in which a solid–fluid phase transition alone may develop. Some indication of the slope of the liquid–gas boundary of stability may be obtained from the Clausius–Clapeyron equation in which we set $\Delta S = L_{vap}/T$ where L_{vap} represents the latent heat of vaporization of the liquid. L_{vap} represents the work done in extracting a molecule from its cage of nearest neighbours in the liquid and transferring it to the vapour phase. Clearly this will depend both upon the coordination—the number of nearest neighbours —and the well depth ε. In most dense liquids the coordination is ~ 8–10 nearest neighbours, in which case it is primarily ε and ΔV which determines the slope of the branch QC. As we see from Figure 5.4(b), ΔV depends quite sensitively upon the form of the attractive branch, in particular upon the index m. It follows that for a given well depth ε the slope increases with increasing m.

At high densities in the vicinity of the solid–fluid phase transition the repulsive aspect of the pair potential will dominate, and for non-rigid molecular cores we obtain the approximate equation of state

$$(P + P_0)V_{\text{eff}} \propto T \tag{5.3}$$

where V_{eff} is an effective temperature-dependent volume, and we neglect the weak temperature dependence of the correction to the pressure P_0 due to the attractive forces. If the diameter of the repulsive core varies as $T^{-1/n}$, the effective volume will vary as $T^{-3/n}$ and equation (5.3) becomes

$$P = -P_0 + bT^{1+3/n} \qquad P_0, b = \text{constants}$$

which is of identical functional form to the empirical *Simon equation* which describes the solid–liquid melting line quite accurately for a wide variety of substances. It is quite clear that the boundary of stability between solid and liquid is governed by the details of the repulsive core and is essentially a *geometric* problem—an aspect which will recur again in our discussions of the phase transition. (Indeed, the geometric aspect of melting was first incorporated in the empirical Lindemann law of melting according to which a solid melts when the mean displacement of an atom from its regular lattice site exceeds 10 per cent of the atomic diameter. The Lindemann law successfully accounts for the melting behaviour of a very wide range of substances. Alternatively and equivalently we may say that a substance melts when its volume exceeds by 30 per cent its volume at 0 K.)

Experimentally the temperature dependence of the argon coexistence curve is found to vary as $T^{1.288}$ which is typical of the liquid inert gases, to $\sim T^3$ for the liquid metals. These results are consistent with $10 < n < 12$ for the inert gases which is in good agreement with the Lennard–Jones type of interaction generally believed to operate in these systems. For metals we might expect the positive ions to repel each other with a screened Coulomb interaction, and the experimental results would suggest $1 < n < 2$.

A Formal Theory of Phase Transitions: the Yang–Lee Condensation

As we emphasized in our discussion of the grand partition function (Chapter 2), the grand canonical form is of more general application than the canonical form and can be used in situations where the number of particles is not fixed as, for example, in the case of a two-phase equilibrium. The number particles n_i, n_j in phases i and j can vary continuously from 0 to N subject, of course, to the condition $n_i + n_j = N$. The grand partition function is

$$\Xi = \sum_{N=0}^{\infty} z^N Z_Q(N)$$

$$= 1 + zZ_Q(1) + z^2 Z_Q(12) + z^3 Z_Q(123) + \dots \tag{5.4}$$

and is seen to be expressed as an infinite sum of 1-, 2-, 3-, ..., N-body configurational canonical partition functions. If the total volume of the assembly is fixed at V, there is evidently an upper limit to the number of particles which may be accommodated, N_{max}, and equation (5.4) consequently reduces to a polynomial in z of degree N_{max}:

$$\Xi(z, V) = 1 + z Z_Q(1) + z^2 Z_Q(12) + z^3 Z_Q(123) + \ldots + z^{N_{max}} Z_Q(1 \ldots N_{max})$$

(5.5)

All the coefficients in the polynomial are by definition positive, and consequently equation (5.5) has no real positive root : its roots are entirely complex and occur in conjugate pairs of the form $(z \pm iy)$. We saw in Chapter 2 that the equation of state is given as

$$\frac{P}{kT} = \frac{1}{V} \ln \Xi(z, V)$$

(5.6)

and the specific density, (volume per particle)$^{-1}$, is

$$\rho = \frac{1}{v} = \frac{1}{V} z \frac{\partial}{\partial z} \ln \Xi(z, V)$$

(5.7)

so that P is an analytic function of ρ in the complex plane which includes the real axis.

Should a root of $\Xi(z, V)$ lie on the real positive axis ($\Xi(z, V) = 0$) then equations (5.6) and (5.7) become singular and presumably represent some kind of singular behaviour in the equation of state.

Provided no roots lie on the real axis the following statements hold :

(i) $P \geqslant 0$, since the coefficients $Z_Q(1 \ldots N)$ for each N are positive

(ii) $\dfrac{V}{N_{max}} \leqslant v < \infty$, from the definition of N_{max}

(iii) $\dfrac{\partial P}{\partial v} \leqslant 0$, since $\dfrac{\partial P}{\partial z} \cdot \dfrac{\partial z}{\partial v} \leqslant 0$

Conditions (i), (ii) and (iii) provide a qualitative description of the equation of state of a finite, single-component homogeneous phase. No unusual behaviour is observed for the simple mathematical reason that none of the roots of $\Xi(z, V)$ lie on the positive real axis, but are distributed in conjugate pairs over the complex plane.

There are N_{max} complex roots to the polynomial $\Xi(z, V)$ since it is of order N_{max}: if we proceed to the limit $V \to \infty$ then the order of the polynomial and the number of roots will increase. The roots will move around in the complex plane and some will converge onto the real axis. The development of homogeneous phases separated by singularities in the equation of state now follows from a consideration of the distribution of zeros in the vicinity of the real axis, and this represents the formal approach of Yang and Lee to the theory of phase transitions.

First we quote without proof two theorems of Yang and Lee relating to the behaviour of the pressure as we pass to the limit $V \to \infty$. For the proof of these theorems the reader is referred to one of the standard statistical mechanical texts.

THEOREM 1. $\text{Lt}_{V \to \infty} P(z, V)$ exists for all $z > 0$ regardless of the shape of the volume V, and is a continuous increasing function of z.

THEOREM 2. If there is a region X of the real positive axis containing no root of $\Xi(z, V)$ for any V, then $\text{Lt}_{V \to \infty} P(z, V)$ converges uniformly to a limit as $V \to \infty$. This limit is a function of z for all z in X.

If the region X_i includes the entire positive real axis as shown in Figure 5.5, then the conditions (i), (ii) and (iii) would still be expected to hold. A possible

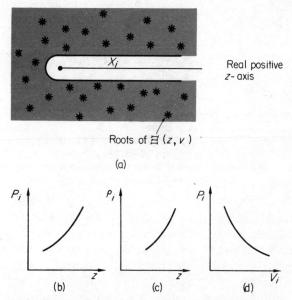

(a)

(b)   (c)   (d)

Figure 5.5 (a) Distribution of roots of the grand partition function in the complex plane. The region X_i of the real positive axis is free of roots. (b) Variation of the pressure of phase i with fugacity, z. (c) Variation of the density of phase i with fugacity, z. (d) Graphical elimination of z-dependence: variation of pressure with volume in phase i

equation of state consistent with the Yang–Lee theorems is shown in Figure 5.5. The z-dependence of P and ρ has been eliminated graphically to yield the equation of state shown in Figure 5.5(d). We see that the absence of any zeros on the real positive axis corresponds to a homogeneous single phase with no

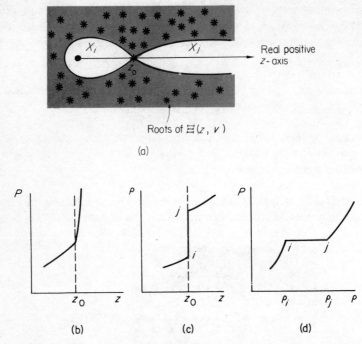

Roots of $\Xi(z, V)$

(a)

(b)   (c)   (d)

Figure 5.6 (a) Distribution of roots in the complex plane in the vicinity of a phase transition $i \leftrightarrow j$. Sections of the real positive axis are isolated by the root at z_0. (b) Variation of pressure with fugacity z in vicinity of z_0. (c) Variation of density with fugacity in vicinity of z_0. (d) Graphical elimination of z-dependence showing the pressure–density isotherm in the vicinity of the phase transition $i \leftrightarrow j$

phase transition. If, however, a root of $\Xi(z, V)$ touches the real axis at z_0 then two regions X_i and X_j are created in which conditions (i), (ii) and (iii) are separately satisfied. These regions evidently correspond to two distinct phases ($z < z_0, z > z_0$), separated by a singularity in the equation of state at z_0 (Figure 5.6a). From Theorem 1, P is evidently a *continuous* function of z, even at z_0 (Figure 5.6b), although its derivative may be discontinuous. At z_0 $\rho(z)$ is discontinuous, and it may be shown that if z_i lies in X_i and z_j lies in X_j where $z_j > z_i$, then $\rho(z_j) > \rho(z_i)$ (Figure 5.6c): the final equation of state, Figure 5.6(d), shows a first-order phase transition.

As the temperature varies the roots will move on the real axis, and if at some temperature the roots cease to converge onto the real axis, then this evidently corresponds to the critical temperature T_c for the phase transition $i \leftrightarrow j$. On the other hand, if two roots merge at some temperature T_0, we then have a triple point at that temperature.

We may therefore relate the occurrence of a phase transition to the fact that a root of $\Xi(z, V)$ has touched the real positive z-axis as we proceed to the limit $V \to \infty$. Of course, the above development does not apply to real systems of finite size: phase transitions are not singularities, but merely finite changes in derivatives. These difficulties are removed if we deal only with infinite systems, but we believe that provided we treat large but finite systems, the predictions of statistical mechanics are confirmed within experimental error. There is no difficulty, however, in constructing a *polynomial* of order 10^{23} with a change of slope sufficiently abrupt as to appear discontinuous. We shall encounter the consequences of these very difficulties when we attempt to study phase transitions in the small finite molecular dynamic and Monte Carlo simulations of phase transitions. Typically such a system consists of only $\sim 10^3$ particles, and in no case is a first-order phase transition with a horizontal transition isotherm or *tie-line* ever observed. These small systems presumably suppress the larger density fluctuations which accompany the phase transition, and the transition is accordingly inhibited.

Another important feature is the essentially geometric one which governs the order N_{max} of the polynomial $\Xi(z, V)$. For a finite system of given volume the number of roots and their convergence onto the real axis is governed almost entirely by geometric considerations: it is interesting that even in a formal theory of phase transitions the geometrical packing aspect of the problem is asserted.

Machine Simulations

The use of computer-simulated assemblies of interacting particles provides the most direct insight into the microstructure and microkinetics of solid and liquid systems and has strongly substantiated the essentially geometric aspect of the melting transition.

The molecular dynamics equation of state of an 870 hard-disc assembly is shown in Figure 5.7. At low to intermediate densities the fluid isotherm shows a monotonic increase until at $PA_0/NkT \sim 7.7$ there is a transition to a denser, crystalline locus representing the solid branch of the isotherm. The solid and fluid branches are not linked by the horizontal tie-line we might expect, but instead by a van der Waals loop, shown in detail in Figure 5.7. In an infinite system we have seen from the Yang–Lee theory of condensation that the two branches of a first-order phase transition should be linked by a horizontal isotherm AB, and the molar free energies g_A, g_B should be identical. Practical considerations restrict molecular dynamic and Monte Carlo simulations to small finite assemblies typically with $N < 1000$ particles. The loop almost certainly derives from the constraint of constant density which is imposed (by virtue of the fixed cell volume) at each density point. The density fluctuations which generally accompany the phase transition are evidently suppressed by the boundaries of the assembly and the reversible exchanges between the two phases are correspondingly inhibited. The effect of this constraint turns out to stabilize the predominant phase: thus the system was either all solid, or, when a rare fluctuation disordered enough of the system, it became entirely fluid. The metastable sections of the isotherm AA', BB' have a real physical significance

134

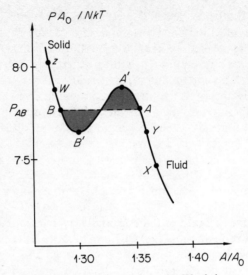

Figure 5.7 Detail of the van der Waals loop and application of the Maxwell equal-areas rule. The tie-line AB is located so as to make the shaded areas equal

in as far as large density fluctuations are inhibited in very small droplets and the metastable regions are physically realizable, though not of course the intrinsically unstable region $A'B'$. There is, therefore, some uncertainty in the location of the tie-line AB, and the Maxwell equal-areas rule is generally used to locate AB such that the shaded areas are equal (Figure 5.7). When one mole of the liquid is changed into vapour the change in the Gibbs free energy is

$$g_A - g_B = \int_{V_A}^{V_B} P \, \mathrm{d}v \tag{5.8}$$

and since we require $g_A = g_B$ at the phase transition it follows that the shaded areas should be equal.

It is instructive to consider the schematic variation of the molar free energy as the pressure is increased (Figure 5.8). At a pressure corresponding to the point X on the fluid branch the density and Gibbs free energy are uniquely defined. At the pressure P_Y there are *two* possible values of ρ and g: thereafter there are *three* possible values and the Gibbs free energy shows the variation with pressure across the van der Waals loop shown in Figure 5.8. If now we follow the locus corresponding to minimum Gibbs free energy we see that the metastable branch of the transformation is not relevant to the phase transition, and the horizontal tie-line evidently corresponds to stable coexistence between two phases of identical and minimum Gibbs free energy. Alder and Wainwright have shown that A and B on the hard-disc isotherm do indeed correspond to states of identical Gibbs free energy.

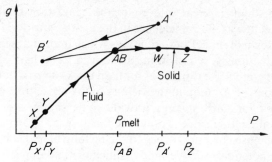

Figure 5.8 Schematic variation of Gibbs free energy with pressure in vicinity of the phase transition shown in Figure 5.7. The tie-line is characterized by a single point AB on the diagram corresponding to continuity of the free energy, though a discontinuity in the derivative $(\partial g/\partial P)_T$. The path $A'B'$ represents the unstable portion of the loop $A'B'$ shown in Figure 5.7

The 'crystallography' of the fluid and solid branches of the isotherm in the vicinity of the phase transition is shown in Figure 5.9: the quasi-crystalline arrangement seems characteristic of the solid locus, whilst the more disordered structure is characteristic of the fluid branch of the hard-disc isotherm. Whilst

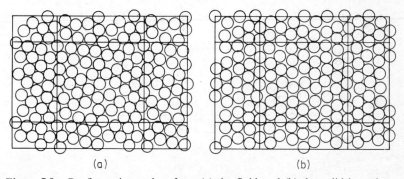

(a)   (b)

Figure 5.9 Configurations taken from (a) the fluid and (b) the solid branches of the hard-disc isotherm

the machine simulations cannot prove rigorously that the phase transition is of first order, they nevertheless give the strongest indication that this is what they would represent for an infinite system. Of course, a two-dimensional assembly of N particles effectively represents a much larger system than the same number of particles confined in a three-dimensional array: the two-dimensional computations do, therefore, adopt an important significance in the computer simulations, and would be expected to demonstrate all the qualitative features of the solid–fluid phase transition.

Nevertheless, there is great interest in the simulation of the hard-sphere isotherm: indeed it was the coincidence of the phase transition for rigid spheres and the breakdown of the KBGY equations which initiated the search amongst the integral equations for transition phenomena. The hard-sphere isotherm shows the same qualitative features as for hard discs—in particular the development of a van der Waals loop in the vicinity of the phase transition. Again, the results obtained for hard spheres provide some support for the geometric interpretation of melting since it occurs in the absence of attractive forces, yet demonstrates a close correspondence to the melting of real substances.

For realistic potentials the thermodynamic functions depend explicitly upon temperature, and in Figure 5.10 we show the three-dimensional 126 K isotherm

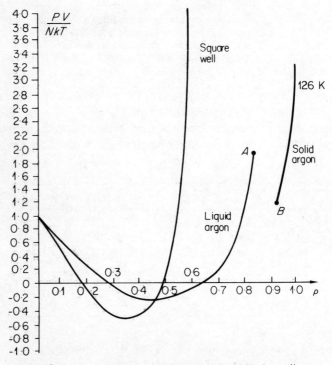

Figure 5.10 Comparison of the simulated 126 K three-dimensional argon isotherm with a square-well simulation. The attractive region of the square-well interaction ensures the same qualitative features in both curves

for argon. The isotherm shows two distinct branches at high density which may be attributed to the solid and liquid phases. The transition is poorly defined as always and the Maxwell equal-area rule is difficult to apply. Nevertheless, the radial distribution functions have been determined at the points A and B (Figure 5.11) and they seem to show a greater structural development on the solid locus with respect to the liquid.

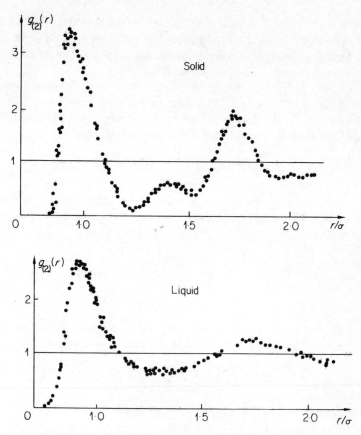

Figure 5.11 The radial distributions taken from the points A and B on the argon simulation. The solid RDF shows slightly more fine detail than the liquid distribution

The KBGY Equations and the Kirkwood Stability Criterion

Quasi-thermodynamic models have been proposed such as that of Kerber, in which an anharmonic solid melts into a Percus–Yevick liquid. The molar Gibbs free energies for each phase are calculated as a function of temperature and their intersection (corresponding to the equality $g_l(P, T) = g_s(P, T)$) locates the melting pressure (see Figure 5.8). The discontinuities in volume and entropy then follow directly from the two equations of state evaluated at that pressure. A number of such calculations at different temperatures enable the solid–liquid coexistence curve to be determined, but the results for argon are not good due primarily to the great sensitivity of the thermodynamic functions to inaccuracies in the PY equation at high densities. Whilst such a direct approach to the melting problem is attractive from a thermodynamic point of view, it lacks the *a priori* objectivity of a single theory which describes *both* the solid and the fluid phases and to that extent evades the central problem—the molecular processes of the phase transition. It is of course in this respect that the KBGY class of equations is unique, ceasing to yield fluid-like solutions beyond some critical density which, for hard-sphere assemblies at least, coincides almost exactly with the molecular dynamics solid–fluid phase transition.

In the vicinity of a phase change the transition may be initiated by a small density fluctuation: Kirkwood has investigated the stability of the BGY equation with respect to small perturbations in the single particle density. From equation (3.29) we may write for the single-particle distribution

$$-\beta^{-1}\rho_{(1)}(\mathbf{1})\nabla \ln \rho_{(1)}(\mathbf{1}) = \int \nabla\Phi(12)\rho_{(2)}(\mathbf{12})\,d2 \qquad (5.9)$$

The following perturbations to the one- and two-particle density distributions are now made, and Kirkwood incorrectly assumes that these may be formed independently:

$$\left.\begin{array}{l} \rho_{(1)}(\mathbf{1}) \to \rho_L[1 + \phi(\mathbf{1})] \\[6pt] \rho_{(2)}(\mathbf{12}) = \rho_{(1)}(\mathbf{1})\rho_{(1)}(\mathbf{2})g_2(\mathbf{12}) \\[6pt] g_{(2)}(\mathbf{12}) \to g_L(12) + \chi(\mathbf{12}) \end{array}\right\} \qquad (5.10)$$

where ϕ and χ are small perturbations to the homogeneous liquid distributions ρ_L and g_L. Linearizing equation (5.9) with respect to ϕ and χ we obtain

$$-\beta^{-1}\nabla\phi(\mathbf{1}) = \rho_L \int \nabla\Phi(12)[g_L(12)\phi(\mathbf{2}) + \chi(\mathbf{12})]\,d2 \qquad (5.11)$$

It is then easy to solve for ϕ by Fourier transformation and use of the convolution theorem:

$$\tilde{\phi}(k) = M(k)/[1 - G(k)] \qquad (5.12)$$

where

$$\left.\begin{array}{l} M(k) = \dfrac{i}{k^2}\int \mathbf{k} \cdot \Delta(\mathbf{x}_1)\exp(-i\mathbf{k}\cdot\mathbf{x}_1)\,d\mathbf{x}_1 \\[12pt] \Delta(\mathbf{x}_1) = \beta\nabla_1\phi(\mathbf{x}_1) + \beta\rho_L \int \nabla_1\Phi(x_{12})\chi(\mathbf{x}_1\mathbf{x}_2)\,d\mathbf{x}_2 \end{array}\right\} \qquad (5.13)$$

and

$$G(k) = \frac{i\beta\rho_L}{k^2}\int \frac{\mathbf{k}\cdot\mathbf{x}}{x}\Phi'(x)g_L(x)\exp(-i\mathbf{k}\cdot\mathbf{x})\,dx$$

Thus, an instability occurs in (5.12) when $G(k) - 1$ vanishes, and this is determined solely by unperturbed quantities. Equation (5.12) becomes singular at certain points on the density–temperature plane, and their locus describes the boundary of stability of the liquid phase.

Equations (5.12) and (5.13) adopt a particularly simple form in the case of hard spheres when $\Phi'(x)g_L(x)$ reduces to a delta function at the atomic diameter,

139

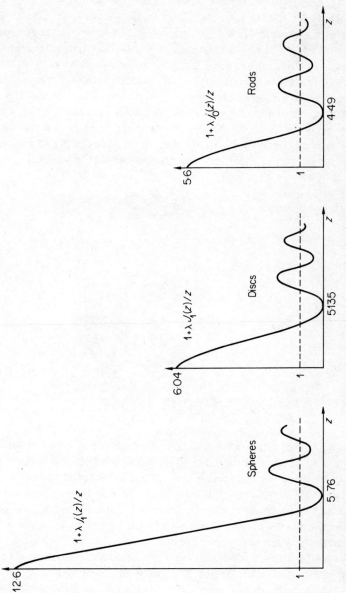

Figure 5.12 Conditions for the development of long-range oscillatory solutions in one, two and three dimensions

$x = \sigma$. $G(k)$ represents the Fourier transform of the force distribution $g_{(2)}\nabla\Phi$ which has an essentially geometric interpretation in the case of rigid-core (hard, spheres, discs, rods) interactions. A density-dependent (i.e. λ-dependent) singularity in equation (5.12) at some critical value of $z_c = (k\sigma)_c$ shows that the single-particle distribution $\rho_{(1)}(1)$ has developed a lonb-range oscillatory structure of wavelength $(2\pi/k_c)$ (Figure 5.12). This arises through the singular development of $\phi(k_c)$. The wavelength of the single-particle functions is seen to be lower in three dimensions $(2\pi/5.76)_{3D}$ than in two and one dimension $(2\pi/5.135)_{2D}$, $(2\pi/4.49)_{1D}$. This may be understood in terms of the intervention of indirect correlative effects in the higher dimensions. The singularities occur in the random dense-packed contact configuration and this 'log-jamming' will evidently occur in three dimensions for a more open packing configuration than in two or one dimensions. In one dimension, for example, this cannot occur much before the simple linear-contact configuration (Figure 5.13a).

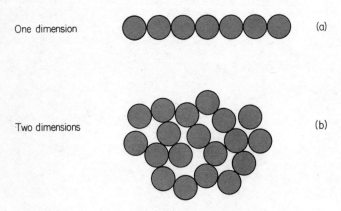

One dimension   (a)

Two dimensions   (b)

Figure 5.13   Contact configurations (a) in one dimension, (b) in two dimensions

Evidently there exists a critical λ above which the system cannot be stable with respect to small perturbations in the single-particle distribution function. We note that this treatment requires no assumption about the triplet distribution function: it was at one time thought the Kirkwood's original stability criterion, the breakdown of the *two-particle* KBGY equation, was dependent upon the superposition approximation and the linearization of the equations.

The single-particle and two-particle distributions are not, of course, independent: this self-consistency is obvious from the hierachical structure of the KBGY class of equations. Nevertheless, in equations (5.10) Kirkwood incorrectly assumes that perturbations in $\rho_{(1)}$ and $\rho_{(2)}$ may be formed independently, which they cannot, and attempts have been made to modify the Kirkwood stability criterion in the light of these shortcomings, but without success.

Qualitatively encouraging though these results are in three dimensions, the Kirkwood stability criterion predicts a hard-disc phase transition at a critical pressure $(P/\rho kT)_K = 4.78$ which is far below the molecular dynamics result $(P/\rho kT)_{MD} = 7.72$ (Figure 5.7). More serious, however, is the prediction of a one-dimensional phase transition which is known not to occur, and the identification of the singularity with the onset of long-range crystalline order must be regarded with some suspicion. If the two-particle KBGY equations were to yield long-range oscillatory solutions beyond the critical density then the physical identification of the singularity would be more straightforward, but it is impossible to obtain meaningful solutions on the 'solid' side of the singularity. This is a major shortcoming since a fully adequate theory should surely be capable of describing the structural and thermodynamic features on both sides of the transition. Of course, the structure of the solid is not spherically symmetric as is the liquid—a subsequent free-energy calculation would be necessary to distinguish between the various crystal structures which could develop. Nevertheless, the relative stability of the hexagonal and cubic structures is so small as to remain unresolved in any approximate theory of phase transitions, and there seems no reason why the spherically symmetric distribution functions should not provide a good, if angle-averaged, structural and thermodynamic description of a hexagonal close-packed solid.

A recent analytic approach to the theory of phase transitions has been given by Weeks, Rice and Kozak. A criterion for the *uniqueness* of the solution to the single-particle Kirkwood integral equation is given, and on this basis the region of the ρT plane is found over which the one-dimensional Kirkwood solution is unique. Multiple solutions of the non-linear equation are associated with instability of the single phase, and thus signal a phase transition. A *bifurcation equation* which can be related to Kirkwood's instability criterion describes the initiation of multiple solutions: the periodic single-particle density falls naturally out of the theory. No phase transition is found for a system of hard spheres, however.

Role of the Direct Correlation

Klein and Green have made an exhaustive study of the HNC equation for Lennard–Jones and hard-sphere fluids. We shall only consider here those aspects of their investigation of immediate relevance to phase transitions.

In the case of the Lennard–Jones fluids Klein and Green obtain the equations of state shown in Figure 5.14 for a variety of reduced temperatures. The agreement with experiment is good at low density and shows a progressive discrepancy with increasing density. In particular, we notice that the $T^* = 1.25$ isotherm has two branches, in agreement with experiment—the low-density branch corresponding to what we might call the gas phase, and the high-density branch the liquid phase. These identifications are confirmed by the radial distributions shown in Figure 5.14 along the $T^* = 1.25$ isotherm. There is no indication of a liquid–solid phase transition.

Of particular interest is the calculation of the isothermal compressibility $kT(\partial\rho/\partial P)_T$ which becomes infinite at a first-order phase transition

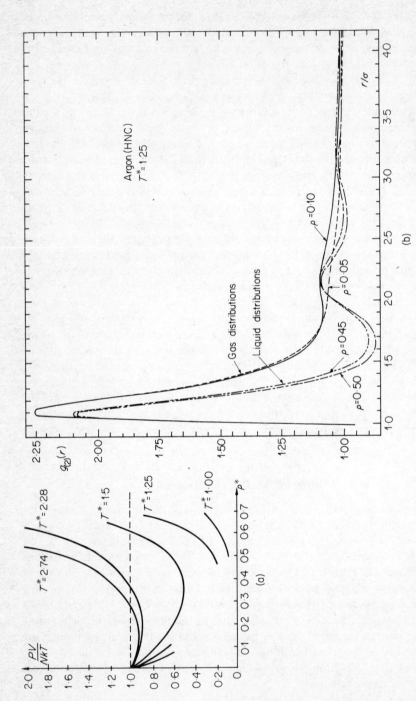

Figure 5.14 (a) The equation of state for liquid argon calculated in the HNC approximation. For reduced temperatures below ~$T^* = 1.4$ there is a region in the $P-\rho$ plane for which there are no solutions. This region is tentatively identified with coexistence. (b) Gas and liquid distributions for liquid argon in the HNC approximations taken from the low- and high-density branches of the subcritical isotherms in (a).

corresponding to the horizontal portion of the PV diagram. We have seen that the isothermal compressibility χ_T is proportional to the long-wavelength limit $(k \to 0)$ of the structure factor:

$$S(0) = kT\rho\chi_T \qquad (5.14)$$

which may be related to the direct correlation through the long-wavelength limit of equation (3.78)

$$S(0) = \frac{1}{1 - \rho c(0)} \qquad (5.15)$$

Singularities in the above equation evidently predict infinite compressibility, and Klein and Green find all their singularities are of this kind of Lennard–Jones fluids. In the case of hard spheres singularities are obtained only for $k > 0$, and this has no immediate physical significance—cetainly not in terms of equation (5.14).

The compressibility curves as a function of density and temperature are shown in Figure 5.15. The intersection of the subcritical isotherms with the density axis

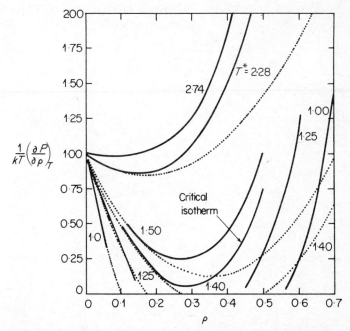

Figure 5.15 Inverse maximum compressibility isotherms for liquid argon in the HNC approximation. Subcritical isotherms intercept the axis indicating infinite compressibilities at those densities. This region of the ρ–T plane is related to coexistence

defines a locus of singularities in the density-temperature plane representing the boundary of stability of the liquid–vapour coexistence. The critical isotherm

144

just touches the density axis, whilst the supercritical isotherms representing a homogeneous gas phase lie entirely above the axis.

Klein and Green obtain the compressibility curves shown in Figure 5.15 on the HNC equation. The zeros of $(kT)^{-1}(\partial P/\partial \rho)_T$ are located on the singularities, and the locus of the singular points defines the liquid–gas coexistence boundary.

It does appear that we have obtained direct evidence of a first-order phase transition in reasonable agreement with experiment. Unfortunately the Klein–Green locus of singularities corresponds to the *limit of metastability* of the phase, that is, to the locus $(\partial P/\partial \rho)_T = 0$. Experimentally, with the exception of the critical point, $(\partial P/\partial \rho)_T \neq 0$ at all points on the coexistence curve. For this reason, the theoretical locus of singularities cannot be identified with the experimental coexistence curve (except at the critical point).

Diagrammatically, the phase transition may be most easily discussed in terms of the direct correlation. First, we recall the temperature-dependent form of the Mayer f-function for realistic interactions. In particular at low temperatures the f-function shows a pronounced positive tail which washes out at higher temperatures when the repulsive core domaines the interaction. Correspondingly, the direct correlation shows a pronounced positive long-range tail, Figure 5.16. The development of the tail increases with density and decreasing temperature. The slope of the isotherm is given by the compressibility

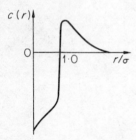

Figure 5.16 The direct correlation for a realistic fluid

equation of state and clearly there will exist combinations of the density and temperature variables for which the integral

$$4\pi\rho \int_0^\infty c(r) r^2 \, \mathrm{d}r = 1 \tag{5.16}$$

(Figure 5.17). Then the compressibility is zero and the isotherm describes the horizontal portion of coexistence. The transition region is evidently governed by the delicate balance between the long-range and short-range components of the integrand, equation (5.16). The r^2 factor emphasizes the details of the long-range tail, whilst suppressing the $r < 1\cdot0$ behaviour. Nevertheless, the core of

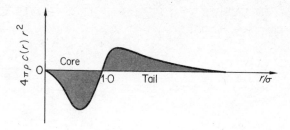

Figure 5.17 The integrand of the compressibility integral. When the shaded area has a net value of unity the derivative $(\partial P/\partial \rho)_T$ becomes zero

the direct correlation is strongly density dependent and but for the short range of coexistence, dominates the form of the equation of state.

The *exact* direct correlation presumably must match the development of the long- and short-range regions of the integrand with density such that $(\partial P/\partial \rho)_T$ remains zero throughout the coexistence. Otherwise the equation of state will show the familiar van der Waals loop as first the long-range tail of the integrand governs the slope, and then ultimately at higher densities the core of the integrand reasserts itself.

At higher temperatures the tail is washed out and the core of the integrand dominates the isotherm. Nevertheless, the vestigial tail is responsible for the initial negative slope of the supercritical isotherms (Figure 5.15).

We can understand the development of the tail of the direct correlation from the singular behaviour of $S(0)$ (equations (5.14) and (5.15)) at the phase transition for then $\rho c(0) \to 1$. This local development of the long-wavelength limit of the direct correlation implies long-range real-space behaviour, and indeed this may be directly attributed to the critical role of the tail in the transition region.

For hard spheres the PY approximation neglects the tail of the direct correlation altogether, and there is no possibility of the system showing a phase transition, given the monotonic development of the core with density. Green has investigated the hard-sphere system in the HNC approximation, and whilst the direct correlation shows a weak positive tail, (Figure 5.18) it is clearly insufficient to initiate transition characteristics in the rigid-sphere isotherm, Figure 4.6. In consequence, Green fails to find any $k = 0$ singularities, although there is a tendency for a singularity to occur at $k \sim 7.0$ corresponding to a wavelength ($= 2\pi/k$) approximately equal to the hard-sphere diameter. Such a singularity was not actually observed however (Figures 4.10a, 4.10b).

It is not difficult to conduct a parallel discussion of the phase transition in terms of the pressure equation of state, but the direct correlation function is rather more fundamental in the theory of fluids and enables us to identify the *diagrammatic* mechanism of the phase transition. We shall consider the strictly diagrammatic approach to the theory of phase transitions in the next section.

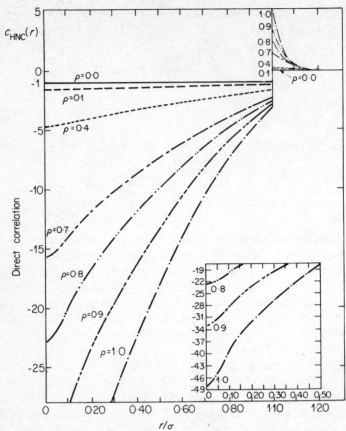

Figure 5.18 The HNC hard-sphere direct correlation as a function of density (cf. Figure 3.21)

Contribution of Small Watermelons

The PY approximation describes the core of the hard-sphere direct correlation function exactly, but neglects the tail completely, and predicts no phase change even beyond physically impossible densities. Thre is no geometrical exclusion within the volume of integration of the field points in the cluster integrals retained in the PY approximation, and the field points may therefore adopt configurations of unphysically high density. The first two members of the set $(B' + E)$ dropped in the PY approximation (⧖ + ⧖) may be written $(1 + f_{34})$⧖, so that to a first approximation the tail may be written $(1 + f_{12})(1 + f_{34})$⧖. This may be represented graphically as where the wiggly bonds represent the factor $(1 + f)$ and effect an *exclusion* on the pair interactions 12 and 34. Thus, the first evidence of geometric packing effects becomes evident in the long-range component of the direct-correlation function.

Croxton has shown that many of the elementary diagrams E, dropped in the PY approximation, may be constructed from the class B' by the application of l Mayer f-bonds $(l = 1, \ldots, \hat{l})$ between the n unbonded field points. Provided we restrict the class of B' diagrams to the so-called *small-watermelon class** which are characterized by having only one field point per chain then it may be shown that the tail

$$(1 + f_{12})(B' + E) \sim (1 + f_{12}) \sum_{n=2}^{\infty} \left\{ 1 + \sum_{l=1}^{\hat{l}} \frac{(-1)^l \hat{l}!}{l!(\hat{l} - l)!} \right\} B'_n \qquad (5.17)$$

where $\hat{l}$ represents the maximum number of interfield-point bonds which may be applied to an nth order B' diagram B'_n. The value of each of the applied bonds is -1 over the range of the hard-sphere f-function: hence the factor $(-1)^l$ in equation (5.17). The other factors account for the number of elementary diagrams, together with their associated weighting factors, which may be generated from a given B'_n diagram by permutation of 1 up to l bonds between the n field points. Provided the range of each of the l applied f-bonds is $\leqslant 1.0$, then the sum over l is non-zero, and in fact is *always* numerically equal to -1. This follows from the binomial expansion of $(1 + x)^n$ with $n = 1$, $x = -1$. In this case then

$$(1 + f_{12})(B' + E) = 0 \qquad \text{for all } l \text{ bonds} \leqslant 1.0 \qquad (5.18)$$

On the other hand, for the range of any or all of the l f-bonds > 1, the E diagrams thus generated must vanish since these elementary diagrams are the *product* of the f-factors. In this case we have

$$(1 + f_{12})(B' + E) \sim (1 + f_{12}) \sum_{n=2}^{\infty} B'_n \qquad \text{for any } l \text{ bonds} > 1.0 \qquad (5.19)$$

From equations (5.18) and (5.19) we therefore conclude

$$(1 + f_{12})(B' + E) \sim (1 + f_{12})\hat{B}' \qquad (5.20)$$

where $\hat{B}'$ represents the evaluation of the small-watermelon class B' *subject to the condition that all field points in that class are separated by unity or more*. This is, of course, a general statement of the geometrical exclusion property of the cluster integrals dropped in the PY approximation. Furthermore, the class $\hat{B}'$ is very small indeed. The stringent requirement that all field points should be separated by an atomic diameter or more, whilst the bonds in the original watermelon diagram remain within a range < 1, rapidly diminishes the contribution of the higher-order diagrams. Indeed, the diagrams show a finite-clustering property in that a fundamental geometrical limit is placed on the

---

* On account of their graphical appearance.

class in the case of hard spheres when twelve particles are in simultaneous interaction: then the exclusion condition has diminished the contribution of the integral to zero. Only those *physically accessible* configurations of field points contribute to the long-range form of the direct correlation, and this may be represented graphically as

$$(1 + f_{12})(B' + E) \sim (1 + f_{12})[\rho^2 \boxtimes + \rho^3 \text{⬠} + \ldots] \qquad (5.21)$$

where the wiggly bond represents the *exclusion* $\tilde{f} = (1 + f)$ operating between field points, which prevents them approaching closer than an atomic diameter (Figure 5.19b). In as far as the diagram $\boxtimes$ completes the diagrammatic contribution to four-particle configurations, we may anticipate that equation (5.21) ensures that the fourth virial coefficient is given exactly.

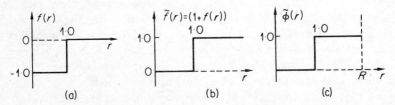

Figure 5.19 (a) The hard-sphere Mayer f-function. (b) The Ree–Hoover function $\tilde{f} = 1 + f$. (c) The function $\tilde{f}$ truncated at R ($> 1\cdot0$) = $\tilde{\phi}$

We need to sum (5.21), but first consider only the exclusion projection of the series: we then have the series of *infinite-range* diagrams (since the wiggly bond $= 1\cdot0$ for $r > 1\cdot0$)

$$\rho^2 \underset{3 \quad 4}{\sim\!\sim} + \rho^3 \underset{3 \quad 4}{\overset{5}{\triangle}} + \rho^4 \underset{3 \quad 4}{\overset{6 \quad 5}{\boxtimes}} + \ldots \qquad (5.22)$$

which may be represented approximately by

$$\rho^2 \underset{3 \quad 4}{\sim\!\sim} + \rho^3 \underset{3 \quad 4}{\overset{5}{\triangle}} + \rho^4 \underset{3 \quad 4}{\overset{6 \quad 5}{\square}} + \ldots \qquad (5.23)$$

i.e. as a netted ring of exclusion bonds. Whilst this series does not *completely* preserve the exclusion characteristics, the field point accessibility remains much the same, and more particularly satisfies a simple convolution relation whose Fourier transform is

$$c_{\text{tail}}(k) \sim \rho^2 \tilde{f}(k)/[1 - \rho \tilde{f}^2(k)] \qquad (5.24)$$

where $\tilde{f}(k)$ is the transform of $\tilde{f}(r)$. If we now artificially *truncate* the range of the $\tilde{f}(r)$ bond at some reduced radius $R > 1$ (Figure 5.19c), we partially recover the finite-ranged series of diagrams, equation (5.21), whilst preserving their exclusion characteristics. Designating this truncated function $\tilde{\phi}(r)$, equation (5.24) becomes

$$c_{\text{tail}}(k) \sim \rho^2 \tilde{\phi}(k)/[1 - \rho\tilde{\phi}^2(k)] \tag{5.25}$$

For hard spheres $\tilde{\phi}(k)$ has a principal maximum at $k \doteq 0$ when $\tilde{\phi}(0) = \frac{4}{3}\pi(R^3 - 1)$, Figure 5.20 (in this treatment $R^3 = 1$ corresponds to the PY approximation). We see that equation (5.25) first becomes singular in the vicinity of $k = 0$

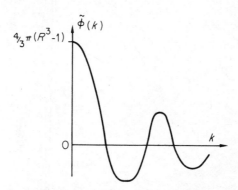

Figure 5.20 The Fourier transformed function $\tilde{\phi}(k)$. This function has the long-wavelength limit $\tilde{\phi}(0) = 4\pi(R^3 - 1)/3$ and develops its maximum at $k = 0$

when $1 = \rho\tilde{\phi}(0)$, and this condition is sensitively dependent upon the range R. Now,

$$S(0) = \rho\beta^{-1}\chi_T = \beta^{-1}(\partial\rho/\partial P)_T = \frac{1}{1 - \rho c(0)} \tag{5.26}$$

Since $c(0)_{\text{PY}}$ becomes progressively more negative with increasing ρ, it follows from equation (5.26) that PY inevitably yields a monotonic compressibility equation of state with no evidence of a solid–fluid phase transition (Figure 5.21). In this small-watermelon approximation, the additional contribution of equation (5.24) results in $\rho c(0) \to 1$ at some density, χ_T becoming infinite at this point. The real-space form of the direct correlation as a function of density for hard spheres is shown in Figure 5.22. The long-range tail shows a dramatic development at the transition density and is responsible for the transition characteristics of the compressibility isotherm as discussed in the previous section (equation (5.16) *et seq.*)

The final equation of state is shown in Figure 5.23(a) and is compared with the hard-sphere molecular dynamics reference isotherm. The tie-line on the theoretical isotherm is located by the usual equal-areas rule, and radial distribution functions from the fluid and solid branches are shown in Figure 5.23(b).

150

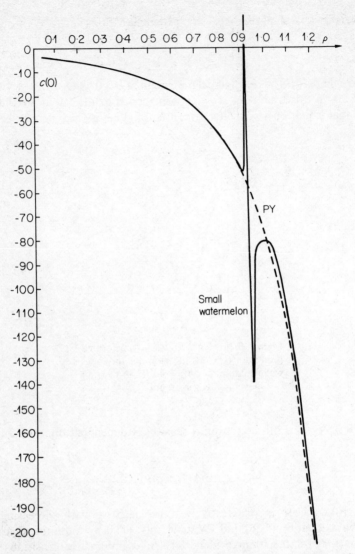

Figure 5.21 Long-wavelength behaviour of the direct correlation in the Percus and small-watermelon approximations. The watermelon approximation develops singular behaviour in the vicinity of the transition density

It may be shown quite easily that in the small-watermelon approximation the potential of mean force Ψ is related to Ψ_{PY} as

$$\frac{\Psi}{kT} = \frac{\Psi_{PY}}{kT} - \hat{B}' \tag{5.27}$$

where $\hat{B}'$ is the long-range tail of $c(r)$. The diagrams $\hat{B}'$ evidently enhance the

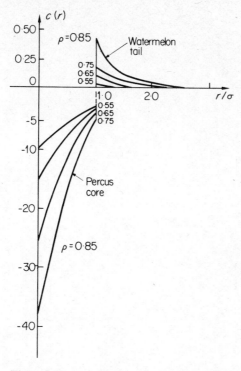

Figure 5.22 Development of the hard-sphere direct correlation with density in the small-watermelon approximation. The tail develops spectacularly in the vicinity of ρ_{trans} whilst the core develops as PY

oscillations in the radial distribution function, which is precisely what is required to correct the PY pressure curve on to the reference isotherm.

The component $-\hat{B}'$ in equation (5.27) represents an additional statistical attraction which develops in the small-watermelon approximation, but not in PY. This additional attraction develops in the assembly for reasons of kinetic shielding amongst the particles. As the density increases so does the kinetic shielding between particles 1 and 2 at small to intermediate separations until at close packing the shielding is complete and the statistical attraction is fully developed. The effect of an attraction, statistical or otherwise, is of course to lower the pressure of the assembly, and this we observe in the rigid-sphere isotherm (Figure 5.23a). A theory such as PY which neglects the geometric interaction amongst the field points allows them unrestricted access and forfeits the effects of shielding. This applies regardless of density, and the system doesn't 'notice' the special geometrical problems arising at the transition density. The watermelon approximation takes explicit account of the geometrical exclusion amongst the field points; the structure-dependent shielding effects are able to

152

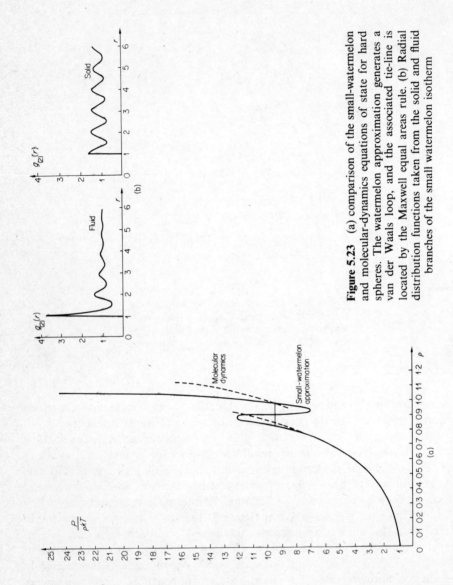

Figure 5.23 (a) comparison of the small-watermelon and molecular-dynamics equations of state for hard spheres. The watermelon approximation generates a van der Waals loop, and the associated tie-line is located by the Maxwell equal areas rule. (b) Radial distribution functions taken from the solid and fluid branches of the small watermelon isotherm

assert themselves, in particular under circumstances when geometrical packing considerations are important, i.e. at ρ_{trans}.

It should be emphasized that it is the *cancellation* or *diagrammatic interference* amongst the watermelon diagrams and their elementary derivatives which is responsible for the features reported here. Their HNC approximation, whilst incorporating the entire watermelon class, may be shown not only to have an incorrect short- and long-range form (equation (3.61)), but in dropping the entire E class, forfeits the exclusion characteristics which seem to be associated with geometric packing.

References

K. Huang, *Statistical Mechanics*, Wiley (1963), Chapter 15.
C. A. Croxton, The Solid–fluid phase transition: the contribution of small watermelons, *J. Phys. C.* 7, 3723 (1974).

CHAPTER 6

The Critical Point

Introduction

At subcritical temperatures $T < T_c$, a gas may be condensed into a liquid by isothermal compression, whilst at supercritical temperatures $T > T_c$ there is no formal distinction between the two fluid phases, either theoretically or experimentally on the basis of discontinuities or higher-order singularities in the equation of state. A qualitative account of condensation and critical phenomena is given by the van der Waals equation of state taken in conjunction with the Maxwell equal-areas rule for subcritical isotherms.

The critical point is located at the apex of the coexistence boundary at which $(\partial P/\partial V)_T = (\partial^2 P/\partial V^2)_T = 0$, and is characterized by singularities in the macroscopic thermodynamic functions, many derivatives becoming infinite or zero. The abnormal features of the fluid in the vicinity of the critical point warrants a separate discussion, and in this chapter we attempt to describe the singular behaviour of the system by expanding the state variables about the critical point, that is, in terms of $\Delta T = |T - T_c|$, $\Delta P = |P - P_c|$ and $\Delta \rho = |\rho - \rho_c|$. As we shall find, the *critical-exponent description* of the critical behaviour of some quantity X as

$$X \propto |T - T_c|^m$$

reveals very important discrepancies between the experimental and 'classical' theoretical exponents m. A classical theory of critical phenomena assumed that the equation of state may be Taylor expanded about the critical point, and this proves to be quite inadequate. Whilst rigorous theoretical results exist only for idealized-lattice gas systems in two and three dimensions—there are few significant theoretical results for realistic systems as yet—the weakness of the classical theory is experimentally conclusive. We shall therefore be primarily concerned with the nature of the departure from classical behaviour, and we shall attempt to establish relations between the critical exponents which must of course exist for any self-consistent theory of critical phenomena.

Experimental measurements become very difficult in this region. Gravitational forces on the sample become important due to the very large compressibility of the system in the vicinity of the critical point. Thermodynamic instabilities may also develop because of the large fluctuations of specific heat and density with temperature—it may take several *days* for the system to finally assume equilibrium; hysteresis and impurity effects present further problems.

Critical Properties of the van der Waals Fluid

In as far as the van der Waals equation of state is qualitatively representative of a large number of simple systems, we determine the critical properties of a van der Waals fluid by Taylor expanding the equation of state about the critical point. The following analysis applies in fact to *any* system for which we can perform Taylor expansions at the critical point. We put

$$\Delta P = P - P_c, \quad \Delta T = T - T_c \quad \text{and} \quad \Delta\rho = \rho - \rho_c$$

and write the equation of state in the vicinity of the critical point as

$$\Delta P = a(T) + b(T)\Delta\rho + c(T)\Delta\rho^2 + d(T)\Delta\rho^3 + \dots \tag{6.1}$$

where $a(T)$, etc. are unknown functions of temperature. At the critical point $\Delta\rho = 0$, $\Delta P = 0$, $\Delta T = 0$ and so we may write

$$a(T) = a_1\Delta T + a_2\Delta T^2 + \dots \tag{6.2}$$

Since the compressibility is infinite at the critical point, i.e. $(\Delta P/\Delta\rho)_{T_c} = 0$

$$b(T) = b_1\Delta T + b_2\Delta T^2 + \dots \tag{6.3}$$

The inflexion in the critical isotherm suggests that the *cubic* term in (6.1) remains at T_c, whilst the parabolic term does not:

$$\left.\begin{aligned} c(T) &= c_1\Delta T + \dots \\ d(T) &= d_0 + d_1\Delta T + \dots \end{aligned}\right\} \tag{6.4}$$

We thus obtain for small $\Delta\rho$ and ΔT the isotherms

$$\boxed{\Delta P \sim a_1\Delta T + b_1\Delta T\Delta\rho + d_0\Delta\rho^3 + \dots} \tag{6.5}$$

It immediately follows from equation (6.5) that along the critical isotherm $(\Delta T = 0)$

$$\boxed{\Delta P \propto (\rho - \rho_0)^\delta} \tag{6.6}$$

where $\delta = 3$. Equation (6.6) describes the critical isotherm in Figure 6.1. Equation (6.5) may be rearranged as follows

$$\frac{\Delta P}{\Delta\rho} - a_1\frac{\Delta T}{\Delta\rho} \sim b_1\Delta T + d_0(\Delta\rho)^2 \tag{6.7}$$

Now, we have

$$(\partial P/\partial\rho)_T \to 0 \quad \text{as} \quad T \to T_c$$

and

$$(\partial P/\partial\rho)_T = \left(\frac{\partial P}{\partial T}\right)_V\left(\frac{\partial T}{\partial\rho}\right)_P \to 0 \quad \text{as } T \to T_c$$

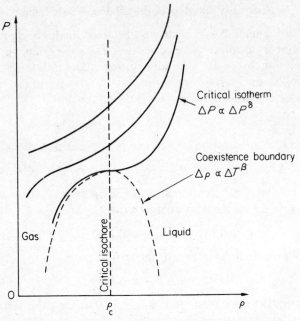

Figure 6.1 Critical features of the P–ρ projection of the equation of state for a realistic single-component system

from which it follows that $(\partial T/\partial \rho)_P \to 0$ as $T \to T_c$ since $(\partial P/\partial T)_V$ tends to a constant value (see for example, Figure 6.1). We immediately have from equation (6.7)

$$\boxed{\Delta \rho \propto (T_c - T)^{\beta}} \qquad (T < T_c) \qquad (6.8)$$

where $\beta = \tfrac{1}{2}$. Equation (6.8) describes the coexistence boundary in Figure 6.1.

A third critical relation shows that the compressibility along the critical isochore $\rho = \rho_c$ diverges as a simple pole. Keeping ΔT constant, and differentiating equation (6.5)

$$\left(\frac{\partial P}{\partial \rho}\right)_T \sim b_1 \Delta T + 3d_0 (\Delta \rho)^2$$

so that the isothermal compressibility χ_T follows as

$$\chi_T = \frac{1}{\rho}\left(\frac{\partial \rho}{\partial P}\right)_T \sim \frac{(b_1 \rho_c)^{-1}}{\Delta T + (3d_0/b_1)\Delta\rho^2} \qquad (6.9)$$

from which

$$\boxed{\chi_T \propto \frac{1}{(\Delta T)^{\gamma}}} \qquad (\rho = \rho_c, T \geqslant T_c) \qquad (6.10)$$

where $\gamma = 1$. Equation (6.10) describes the *reciprocal* of the slope of the isotherm along the critical isochore for $T \geqslant T_c$.

Finally, the isobaric and isochoric specific heats C_p and C_v may be shown through the Maxwell relations to be related directly to the isothermal and adiabatic compressibilities, rspectively:

$$\left. \begin{array}{l} C_p = T\left(\dfrac{\partial S}{\partial T}\right)_P \sim \chi_T T V \left(\dfrac{\partial P}{\partial T}\right)_V^2 \sim (\Delta T)^{-\gamma} \\[4mm] C_v = T\left(\dfrac{\partial S}{\partial T}\right)_V \sim \chi_S T V \left(\dfrac{\partial P}{\partial T}\right)_V^2 \end{array} \right\} \tag{6.11}$$

We have already observed that $(\partial P / \partial T)_V$ tends to a constant value in the vicinity of the critical point and we therefore anticipate that the specific heats vary as the compressibilities. C_p therefore diverges as a simple pole at the critical point (see equation (6.10)) whilst classically C_v would be expected to rise to a *maximum* along the critical isochore, and then fall discontinuously as T increases through T_c:

$$\boxed{C_v \sim C_c^{\pm} - D^{\pm}|T - T_c|} \qquad \begin{array}{l} T \gtrless T_c \\[3mm] \end{array} \tag{6.12}$$

$$C_c^{-} - C_c^{+} = \Delta C > 0$$

where the $+\,(-)$ signs apply for $T > (<) T_c$. To avoid prejudicing the conclusions, we may express the temperature dependence of the specific heat in a way which admits to the possibility of singular behaviour:

$$C_v \sim \frac{A^{\pm}}{\alpha^{\pm}} \left[\left(\frac{T}{T_c} \right)^{\pm 1} - 1 \right]^{\alpha_{\pm}} + B_{\pm} \tag{6.13}$$

so that classically $\alpha = 1$.

The results appearing in equations (6.6), (6.8), (6.10) and (6.13) are expressed in critical-exponent form, and for a classical system (i.e. one in which we may form Taylor expansions about the critical point) we have the results given in Table 6.1. Two other critical-exponent relations describe the variation of the surface tension

$$T \sim |T - T_c|^{\mu} \tag{6.14}$$

and the correlation length

$$\kappa \sim |T - T_c|^{+\nu} \tag{6.15}$$

in the critical fluid.

158

Table 6.1. Classical critical exponents

α	β	γ	δ	μ	ν
1	$\frac{1}{2}$	1	3	$\frac{3}{2}$	$\frac{1}{2}$

Interrelations Between the Critical Exponents

The critical exponent theory assumes that all physical quantities X approach the critical point as

$$X \propto |T - T_c|^m$$

so that

$$m = \underset{T \to T_c}{\text{Lt}} \left\{ \frac{\ln X}{\ln |T - T_c|} \right\} \tag{6.16}$$

and similarly for the other variables $\Delta\rho$, etc. It may be that the critical exponents differ on either side of the critical point so that $m^-(T < T_c) \neq m^+(T > T_c)$. We must therefore admit to the possibility of *two* values for each exponent for those physical quantities which exist on both sides of the critical point ($\Delta\rho$, for example, does not exist for $T > T_c$ and so we need only define β: δ is only defined for the critical isotherm $T = T_c$). The various critical exponents are, of course, related through the equation of state, and it should be possible to establish various equalities, or at least inequalities, between the exponents. These should provide some guide for subsequent theoretical development, and enable us to check the interrelations between the experimental data.

Provided these critical exponent laws are valid, it has been shown on the basis of reasonable thermodynamic assumptions for example that

$$\alpha^- + 2\beta + \gamma^- \geqslant 2 \quad \text{(Rushbrooke)} \tag{6.17}$$

and

$$\alpha^- + \beta(\delta + 1) \geqslant 2 \quad \text{(Griffiths)} \tag{6.18}$$

under the conditions $(\partial^2 P/\partial T^2) \geqslant 0$ on the subcritical region of the critical isochore.

Widom has suggested an *ad hoc* equation of state specifically designed to describe the experimental data, and he finds for his model

$$\delta = 1 + \frac{\gamma}{\beta} \quad \text{(Widom)} \tag{6.19}$$

The slightly less stringent rearrangement has also been proposed

$$\gamma^- \geqslant \beta(\delta - 1) \quad \text{(Egelstaff and Ring)} \tag{6.20}$$

Experimental Determination of the Critical Exponents

(i) The Coexistence Curve

Classically, if the vicinity of the critical point is of parabolic form, $\beta = 0.5$. Analysis of experimental data in the near-critical region however suggests a critical exponent $\beta = 0.345 \pm 0.015$ for a wide range of insulating fluids. However, the experimental difficulties involved in the immediate vicinity of the critical point are considerable, and have been outlined above. The closest approaches to the critical point have been made for Xe and CO_2 $(\Delta T/T_c = 10^{-5})$ but the experimental difficulties make these results relatively less reliable. It may be that the classical value of the critical exponent would be attained in the region $\Delta T/T_c < 10^{-5}$, but this appears unlikely. Guggenheim showed some time ago on the basis of a corresponding states analysis that a large number of non-conducting fluids closely follow a coexistence curve characterized by a critical exponent of $\beta = 0.33$ (Figures 6.2, 6.3).

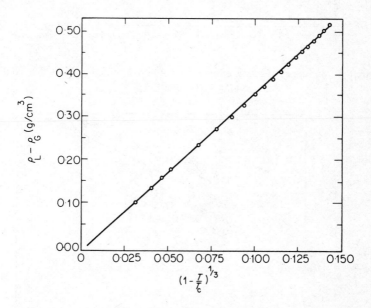

Figure 6.2 Coexistence curve for Xe. The difference in density $\rho_L - \rho_G$ is plotted as a function of the $\frac{1}{3}$ power of the temperature difference from the critical point

A log–log plot for a few alkali metals is shown in Figure 6.4: in this case the *slope* yields a critical exponent in the range 0.42–0.45, which is much closer to the classical value. It is not yet clear whether the long-range features of the pair potential are responsible for the differences between the insulating and conducting fluids.

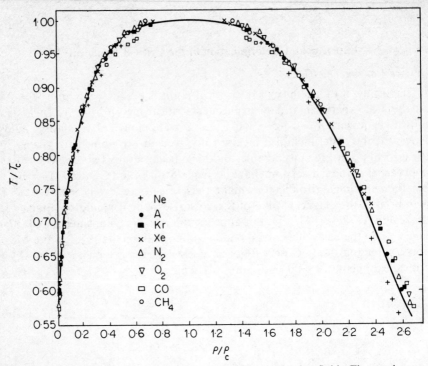

Figure 6.3 Summary of coexistence curves for non-conducting fluids. The continuous curve corresponds to $\beta = \frac{1}{3}$

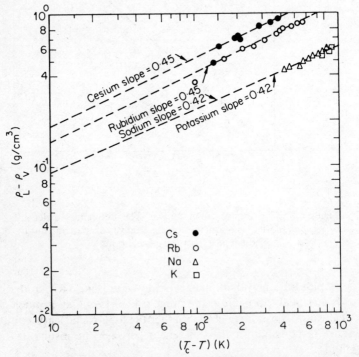

Figure 6.4 Coexistence curves for the alkali metals. The slope of this log–log plot yields the critical exponent β

(ii) *The Specific Heat*

Classically we should expect the specific heat C_v along the critical isochore to rise to a maximum at T_c, with no evidence of a singularity. The experimental work of Bagatskii, Voronel' and Gusak on argon (Figure 6.5) suggests a *logarithmic singularity* from the approximately linear form. Thus the specific

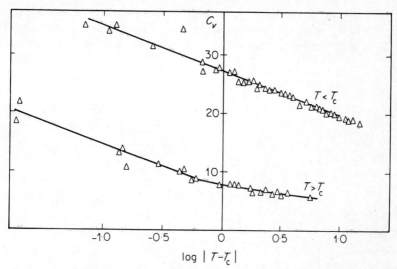

Figure 6.5 Variation of C_v for argon near the critical point along the critical isochore. The linearity of the curve indicates a logarithmic singularity in the specific heat

heat tends to infinity as $T \rightarrow T_c$, and a logarithmic singularity suggests that the critical exponent $\alpha \rightarrow 0$ in equation (6.13). The definite curvature in the data for $T > T_c$ suggests that the true value of α^+ might be somewhat greater than 0·1 whilst below T_c α^- might be less. Results on oxygen yield similar values. Recent theoretical work by Domb *et al.* for simple-lattice gas models suggest $\alpha = 0·125$: no significant predictions are available for more realistic systems. Nevertheless, the weaknesses of the classical model are clearly apparent.

The velocity of sound at low frequencies is of course related to the adiabatic compressibility χ_S, and hence its variation in the vicinity of the specific heat should diverge in the same way as C_v (equation (6.11)—that is, with a critical exponent α^-. A logarithmic divergence is consistent with the experimental data, and again suggests $\alpha^- < 0·1$.

(iii) *The Critical Isotherm*

Classically, the critical isotherm is a cubic (equation (6.6)), and the critical exponent is therefore $\delta = 3$ which differs significantly from the experimental value of $\sim 4·1$ for the systems H_2, CO_2 and Xe. The Griffiths relation

$$\alpha^- + \beta(\delta + 1) \geqslant 2 \tag{6.21}$$

is evidently not satisfied by the experimental data. Xenon, for example, gives

$$0{\cdot}1 + 0{\cdot}34(4{\cdot}1 + 1) \sim 1{\cdot}8 \qquad (6.22)$$

and which of the critical exponents is to blame is not clear—arguments have been advanced against the experimental values of both α^- and δ.

(iv) The Isothermal Compressibility, χ_T

Classically χ_T should diverge at the critical point as $(T - T_c)^{-1}$ ($\gamma = 1$, equation (6.10)) along the critical isochore, but from the experimental data of Habgood and Schneider for xenon shown in Figure 6.6 (note, the *inverse*

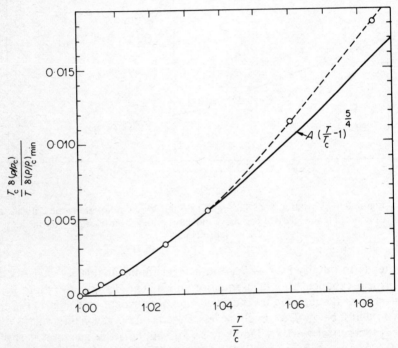

Figure 6.6 Variation of the inverse isothermal compressibility with temperature in the vicinity of the critical point.

compressibility), a critical exponent $\gamma \sim 5/4$ seems more appropriate, at least in the vicinity of T_c. The fit close to the critical point cannot be considered very significant, however, since the experimental measurements become very difficult: nevertheless, there is little doubt that the classical model fails to describe the divergence of χ_T as $T \to T_c$, and does seem to indicate that $\gamma > 1{\cdot}1$.

Unequivocal theoretical calculations of γ exist for idealized-lattice gas assemblies, and the indications are $\gamma = 1{\cdot}75$ (two dimensions) and $1{\cdot}25$ (three dimensions), both of which represent large deviations from the classical prediction of unity.

We may now test the Rushbrooke inequality

$$\alpha^- + 2\beta + \gamma^- \geqslant 2 \qquad (6.23)$$

The experimental evidence is that $0 \cdot 1 > \alpha^- \geqslant 0$, $\beta \sim 0 \cdot 345$: consequently we require

$$\gamma^- \geqslant 1 \cdot 31 \qquad (6.24)$$

if the Rushbrooke inequality is to be satisfied, and this of course differs appreciably from the classical result ($\gamma = 1$).

As far as the three-dimensional-lattice gas model are concerned we have $\beta = 0 \cdot 3125$ and $\gamma^- = 1 \cdot 25$, whereupon we conclude $\alpha^- \sim 0 \cdot 125$. Conversely for a logarithmic specific heat ($\alpha = 0$) we should require $\gamma^- \sim 1 \cdot 35$. It is not yet possible to decide between these alternatives.

We should expect C_p to diverge as the isothermal compressibility (equation (6.11)) and would expect the same value of the critical exponent γ to apply. Unfortunately the experiments are particularly difficult to perform, and although some measurements of C_p have been made for CO_2, the experimental errors are such that the 'critical infinity' is smoothed over, and any estimate of γ on this basis is correspondingly unreliable.

Structural Behaviour Near the Critical Point

The divergence of the isothermal compressibility χ_T at the critical points as $|T - T_c|^{-\gamma}$ implies a similar critical divergence of the long-wavelength structure factor

$$S(0) = \rho\beta^{-1}\chi_T \quad (\beta^{-1} = kT) \qquad (6.25)$$

This kind of divergence is observed experimentally, although as in the case of the 'classical' critical-exponent relations for the thermodynamic quantities, the agreement is only qualitative, and is shown schematically in Figure 6.7. The long-wavelength development in the structure factor implies the existence of long-wavelength fluctuations ($\lambda \sim 5000$ Å) in the near-critical system, and it is of course these fluctuations that are responsible for the critical opalescence or optical scattering observed in the vicinity of T_c. We may understand the rapid formation and disruption of 'flickering' clusters of $\sim 10^3 - 10^4$ molecules from the locus of equilibrium separation, shown by the dotted curve in Figure 6.8. There will be a distribution of molecular energies about the mean, and, as we see from Figure 6.8, this implies large fluctuations in the equilibrium separation near the critical point. Thus, large clusters of tenuously associated particles will continuously form and disrupt—clusters of such a dimension that they may scatter at optical wavelengths.

Now, in k-space we have (equation (3.81))

$$\rho\tilde{c}(k) = 4\pi\rho \int_0^\infty c(r)\frac{\sin kr}{kr}r^2 \, dr \qquad (6.26)$$

164

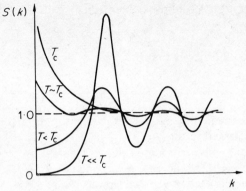

Figure 6.7 Development of the long-wavelength region of the structure factor in the vicinity of the critical point. The long-wavelength limit is proportional to the compressibility and is seen to be low at liquid densities and high in the critical region

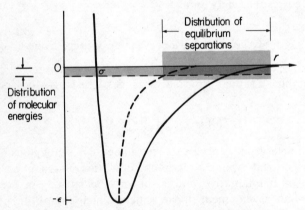

Figure 6.8 Effective region of the pair potential for critical phenomena

for which $\sin kr/kr$ may be expanded at small kr as

$$\sim 4\pi\rho \int_0^\infty c(r)\left[1 - \frac{(kr)^2}{3!} + \ldots\right]r^2\,\mathrm{d}r$$
$$= \rho\tilde{c}(0) - R^2 k^2 + \ldots \tag{6.27}$$

where

$$R^2 = \tfrac{1}{6}\rho \int_0^\infty r^2 c(r)\,\mathrm{d}r \tag{6.28}$$

and we assume this second moment exists, i.e. that $c(r)$ decays faster than r^{-2}.

From equation (6.27) we immediately have from equation (3.78),

$$\frac{1}{\chi_T(k)} \sim 1 - \rho\tilde{c}(0) + R^2k^2 \tag{6.29}$$

i.e.

$$\chi_T(k) = \beta\rho^{-1}S(k) \sim \frac{R^{-2}}{\kappa^2 + k^2} \quad (k^2 \to 0) \tag{6.30}$$

where κ has the dimension of an inverse length and is defined by

$$\kappa^2 = [1 - \rho\tilde{c}(0)]/R^2$$

Fourier inversion of the simple Lorentzian scattering curve (6.30) (Figure 6.9a)

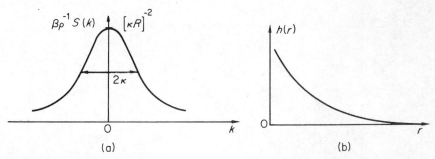

Figure 6.9 (a) Long-wavelength behaviour of the scattering function. (b) Long-range
form of the total correlation in the vicinity of the critical point

shows that the behaviour of the total correlation function for large r in three
dimensions is given as

$$h(r) \sim \frac{1}{4\pi\rho R^2}\frac{e^{-\kappa r}}{r} \quad (r \to \infty) \tag{6.31}$$

In other words, the correlations decay exponentially with a range κ^{-1}, and
from (6.29) the isothermal compressibility follows as

$$\chi_T = \frac{\beta}{\rho R^2\kappa^2} \tag{6.32}$$

Its divergence implies $k(T) \to 0$ as $T \to T_c$ so that (6.31) reduces to

$$h(r) \propto \frac{1}{r} \quad \text{at } T_c$$

From (6.32), knowing the classical divergence of χ_T at the critical point we
conclude that the critical exponent of κ (see equation (6.15) is $v = \frac{1}{2}\gamma = \frac{1}{2}$
(Table 6.1).

From equation (6.30) we would anticipate the family of linear curves shown in Figure 6.10 with the temperature-dependent intercept $\chi_T(0) = R^2\kappa^2(T)$, the critical curve passing through the origin. There is seen to be a pronounced 'non-classical' deviation at small scattering angles indicating some misrepresentation in the assumed long-range structure in the fluid.

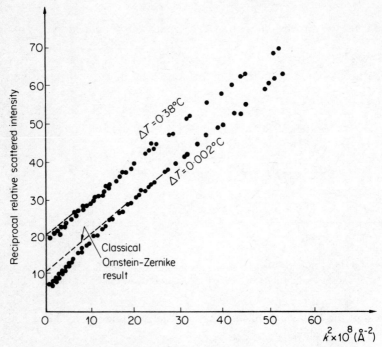

Figure 6.10 Schematic variation of the inverse critical scattering expected in view of the limitations of the classical theory. The 'non-classical' deviation at small scattering angles indicates some misrepresentation in the assumed long-range structure in the fluid

Fisher, in the spirit of the original Ornstein–Zernike expression for the total correlation (6.31) writes more generally

$$h(r) \sim \frac{1}{4\pi\rho R^2} \frac{e^{-\kappa r}}{r^{1+\eta}} \tag{6.33}$$

again for three dimensions. In k-space, omitting constants of proportionality, (6.33) gives the non-Lorentzian critical scattering formula (cf. (6.30))

$$\chi_T \sim \frac{R^{\eta-2}}{(\kappa^2 + k^2)^{1-\eta/2}} \quad (k^2 \to 0) \tag{6.34}$$

which reduces to the classical Ornstein–Zernike result when $\eta = 0$. η is a small constant assumed to lie between 0 and 1, and is regarded as one of the critical indices. The critical indices are now related by

$$\gamma = (2 - \eta)\nu \tag{6.35}$$

Further theoretical analysis suggests that the index $\eta \leqslant 0.1$, and it is difficult to distinguish this from the classical result ($\eta = 0$), though it is clear that there are deviations from classical behaviour consistent with $\eta > 0$.

Critical Surface Tension

Classically, the critical exponent μ in the expression for the surface tension $\propto (T_c - T)^\mu$ is $\mu = \frac{3}{2}$. On the basis of an analysis of the experimental data for a number of insulating fluids, Guggenheim finds $\mu \sim 1.22$ which differs significantly from the classical van der Waals value. Widom has recently reformulated the theory of the critical liquid surface accounting both for the nature of the density transition and the density fluctuations which become significant in the vicinity of the critical point. Widom obtains the consistency relation

$$2\beta + \gamma^- = \mu + \nu \tag{6.36}$$

and taking $\mu = 1.22$, $\beta = 0.345$, $\gamma^- = 1.1$ we recover $\nu \sim 0.57$. Again, from (6.35), we conclude $\eta \sim 0.07$.

Conducting Fluids

It is too early to account for the experimental difference between the critical exponents of insulating and conducting fluids, but in Table 6.2 we compare the two sets of data, and it does appear that there are significant differences.

Table 6.2. Critical exponents satisfying the Rushbrooke–Griffiths consistency relations *within experimental error* (Egelstaff and Ring)

	Classical	Inert gases	Liquid metals
$\alpha^\pm$	1	0.10	0.02
β	$\frac{1}{2}$	0.36	0.44
$\gamma^\pm$	1	1.18	1.10
δ	3	4.28	3.50
η	0	0.15	?
μ	$\frac{3}{2}$	1.26	?
ν	$\frac{1}{2}$	0.64	?

Whether this is to be ascribed to the range or nature of the liquid-metal potential is not known, and given the shortage of liquid-metal data and the difficulties of measurement, it may be that the differences are not as large as indicated. Nevertheless, it must be tentatively assumed that the fundamentally distinct nature of the liquid-metal and inert-gas interactions underlies the distinction.

The Ising Model and Onsager's Solution

The temperature-dependence of the ferromagnetic properties of iron shows many features analogous to the critical behaviour of fluids. Weiss suggested that ferromagnets are characterized by a large internal magnetic field, and proposed that the intensity of magnetization M is related to an external field H through a magnetic 'equation of state':

$$\frac{M}{N\mu} = L\left\{\frac{\mu}{kT}(H + aM)\right\} \tag{6.37}$$

(μ = atomic magnetic moment, $L\{x\}$ = Langevin function = $\coth x - 1/x$). Below some critical temperature T_c, the Curie temperature, the ferromagnet develops a spontaneous magnetization M_0, whilst the magnetic susceptibility becomes infinite as T_c is approached from the high-temperature side. The detailed critical behaviour of this model exhibits precisely *analogous* relations, including the same critical exponents, as the van der Waals model (equations (6.6)–(6.12)).

Not surprisingly a single model, the Ising model, can be used as an approximation to all these critical systems, and it is of great theoretical interest to examine the critical properties of such an idealization—to see in particular whether it provides a more satisfactory description of the experimental data than the classical model, and whether the critical exponents satisfy the various thermodynamic constraints governing their interrelation.

The Ising model allows each spin on a regular crystalline lattice to take on one of two orientations, either parallel or antiparallel to an external magnetic field. Such a model may be extended to give a lattice-gas description of the liquid-vapour equilibrium by identifying one spin orientation with a particle and the other with a hole.

Onsager investigated the critical behaviour of the two-dimensional Ising model, and the results differed completely from the classical theories of Weiss and van der Waals. The results were also in apparent disagreement with experimental observation. Onsager's result provided a challenge to experimentalists and theoreticians alike, for how should we interpret the discrepancy? Did Onsager's solution mean, for example, that the van der Waals and Weiss theories were now completely discredited, despite their satisfactory description of non-critical behaviour? In fact it turns out that for very-long-range attractive interactions with hard cores the van der Waals description coincides with the Onsager solution in one dimension. Again, to what should we attribute the discrepancy between Onsager's solution and experiment? Is it a fundamental inadequacy in the lattice-gas description, or is it the inapplicability of Onsager's essentially two-dimensional solution to three-dimensional systems? It would be a great help if we knew the critical behaviour of a three-dimensional lattice gas, but the mathematical difficulties are so formidable that an exact solution appears remote. Nevertheless, *series* approximations to the three-dimensional Ising system are available and seem to settle down sufficiently to indicate that it is primarily the two-dimensional nature of Onsager's solution which is responsible for the discrepancy with experiment.

It therefore appears that the three-dimensional Ising model can be usefully identified with interactions in real physical systems, and on this basis the hope is that an understanding of critical behaviour will be achieved, together with a more substantial basis for the critical exponent relations outlined earlier in this chapter.

References

P. A. Egelstaff, *An Introduction to the Liquid State*, Academic Press (1967), Chapter 15.

P. A. Egelstaff and J. W. Ring, in *Physics of Simple Liquids* (Eds. Temperley, Rowlinson and Rushbrooke), North-Holland Publishing Co. (1968), Chapter 7.

M. E. Fisher, *J. Math. Phys.*, **5**, 944 (1964).

CHAPTER 7

The Liquid Surface

Introduction

The anisotropy of the liquid surface implies anisotropy of the pair distribution in the transition zone: the spherically symmetric scalar bulk distributions of interparticle separation $g_{(2)}(r_{12}) \rightarrow g_{(2)}(\mathbf{r}_{12}, z_1)$, depending both upon the vector $\mathbf{r}_{12}$ and the location of the origin particle z_1 relative to the liquid surface. This is shown schematically in Figure 7.1. The single-particle distribution $g_{(1)}$ is

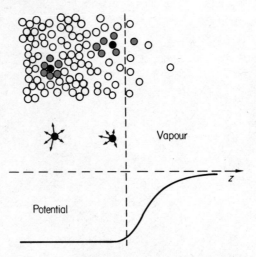

Figure 7.1 Modification of surface structure and atomic coordination in the constraining field at the free liquid surface

of little interest in the bulk liquid, but with $g_{(2)}(\mathbf{r}_{12}, z_1)$, to which it is self-consistently related, assumes central importance in the structural description of the transition zone. Anisotropy in the structure implies anisotropy in the structure-dependent thermodynamic quantities such as pressure, energy and entropy, and it is the task of a statistical mechanical theory of the liquid surface to account for both the structure and principal thermodynamic functions of the liquid surface, such as the surface tension and surface energy. These functions are defined as *excess* quantities, representing the modification of the bulk quantity by the introduction of a free surface. Thus, the forces on a representative molecule become anisotropic and directed into the bulk liquid as it approaches the surface, and work has to be done against the field of the bulk particles to

170

bring it to the surface: a *constraining field* develops at the free surface (Figure 7.1). Thus an excess or *surface energy* develops per unit area due to the introduction of the surface—that is, an excess of the real two-phase system over a hypothetical reference system in which the bulk properties are constant up to a dividing surface separating the bulk vapour from the bulk liquid.

Similarly, we may speak of other excess quantities such as the excess free energy per unit area (surface tension, γ), and the excess entropy per unit area $(-\partial\gamma/\partial T)$ developed at the liquid surface.

Before considering these quantities in any detail we have to specify a coordinate frame in which we can locate a hypothetical boundary between the bulk liquid and vapour phases. Many of the thermodynamic functions of the surface are sensitively dependent on the location of this dividing surface, the surface excess energy for example. If we locate the boundary of the hypothetical two-phase system as in A (Figure 7.2a) then clearly we are calculating the

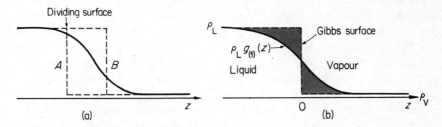

Figure 7.2 (a) Two locations of the hypothetical dividing surface. (b) Location of the Gibbs dividing surface chosen to make the shaded areas equal

excess energy of the real profile over that of an essentially isotropic vapour phase. This will be quite different to the location B in which the excess of the real profile over the isotropic *liquid* phase is determined. Gibbs has shown that the correct location of the dividing zone, and hence the origin of coordinates, is on a plane chosen such that the shaded areas in Figure 7.2(b) are equal. This corresponds to the location of the origin of coordinates such that

$$\int_{-\infty}^{0} [\rho_{\mathrm{L}} - \rho_{\mathrm{L}}g_{(1)}(z)]\,\mathrm{d}z = \int_{0}^{\infty} [\rho_{\mathrm{L}}g_{(1)}(z) - \rho_{\mathrm{V}}]\,\mathrm{d}z \qquad (7.1)$$

i.e. the superficial excess density of matter vanishes.

The surface tension γ is independent of the location of the origin of coordinates, but the excess entropy $(= -\partial\gamma/\partial T)$ is not. Indeed, since the surface tension is the excess free energy per unit area

$$\gamma = \Delta F = \Delta U - T\Delta S \qquad (7.2)$$

only for the location of the origin of coordinates on the Gibbs dividing surface

172

is the surface excess energy $\Delta U = u_s$ related to γ by the thermodynamic relationship

$$\frac{d(\gamma/T)}{dT} = -\frac{u_s}{T^2} \qquad (7.3)$$

and the surface excess entropy given by

$$\frac{d\gamma(T)}{dT} = -s_s \qquad (7.4)$$

Under these circumstances the excess Helmholtz free energy per unit area and the surface tension become identical.

A number of empirical relations between surface tension and other physical properties, such as latent heat of vaporization, have been proposed with varying degrees of success.

We may make a rough identification between the surface excess energy and the work done in reducing the number of nearest neighbours by about 50 per cent in bringing a molecule up to the surface. The work done in rupturing half the molecular bonds is $\sim 0.5 L_{vap}$ where L_{vap} is the latent heat of vaporization per particle. The entropy per particle will change from a value characteristic of the bulk S_L to a value more appropriate to the vapour S_V ($S_L < S_V$), whereupon $T\Delta S \sim \alpha L_{vap}$ where α is a coefficient of proportionality. From equation (7.2) there is evidently a simple linear relationship between γ and L_{vap}, and such a correlation is shown in Figure 7.3 for a number of liquid metals.

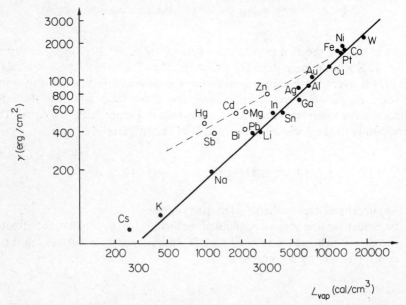

Figure 7.3 Correlation between surface tension and latent heat of evaporation for a number of liquid metals

A Formal Theory of Surface Tension

The pressure at a point in the bulk isotropic fluid is given in terms of molecular quantities as (equation (3.20))

$$P = \rho kT - \frac{\rho^2}{6} \int g_{(2)}(r)\frac{d\Phi(r)}{dr} r \, dr \qquad (7.5)$$

Anisotropy in the pair distribution function at the liquid surface results in a *pressure tensor* appropriate to the local value of the single-particle density $\rho(z)$. Thus, at any point z in the vicinity of the transition zone we may speak of a *normal* pressure component $P_\perp$ perpendicular to the surface, and a *tangential* component $P_\parallel$ parallel to the surface. $P_\perp$ and $P_\parallel$ will differ across the transition zone whilst in either of the isotropic bulk phases they will of course be identical.

For a mechanically stable free surface the component $P_\perp$ must of course be constant, and therefore equal to the equilibrium system pressure P.

Consider a column of fluid of unit cross-sectional area which extends from the bulk liquid across the transition zone into the bulk vapour (Figure 7.4).

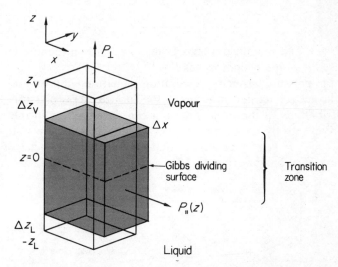

Figure 7.4 Geometry for determination of the mechanical definition of surface tension

Now suppose we perform two isothermal reversible processes:

(1) extend the volume a distance Δx,
(2) compress the volume in the liquid and vapour phases by $\Delta z_L - \Delta z_V$ respectively.

This is done in such a way that, at constant temperature, the final volume and pressure are identical to their initial value, the only difference being an increase in the cross-sectional area by an amount Δx. The total work done in these

174

two processes is

$$-(\Delta w_1 + \Delta w_2) = \gamma(T)\Delta x$$

where

$$\Delta w_1 = \int_{-z_L}^{z_V} P_{\parallel}(z)\Delta x \, dz$$

$$\Delta w_2 = -P_{\perp}[\Delta z_V + \Delta z_L]$$

Since the initial and final volumes are to be identical we have from Figure 7.4

$$(\Delta z_L + \Delta z_V) = (z_L + z_V)\Delta x$$

Thus

$$\Delta w_2 = -P_{\perp}(z_L + z_V)\Delta x = -\int_{-z_L}^{z_V} P_{\perp}\Delta x \, dz$$

So that

$$\gamma(T) = \int_{-\infty}^{\infty} [P - P_{\parallel}(z)] \, dz \qquad (7.6)$$

where $P_{\perp} \equiv P$ and the integral runs from $\pm \infty$, since in the interior of either bulk phase $P_{\parallel}(z)$ rapidly approaches P.

From the pressure-tensor representation (equation (7.6)) we may evidently express the surface tension in terms of the surface excess pressure. It is clear that contributions to the surface tension can develop only in the regions of structural anisotropy, as we might expect of an excess quantity. If γ is to be positive then $P > P_{\parallel}(z)$ generally, so that $P_{\parallel}(z)$ has the character of a *tension* across the transition zone (Figure 7.5).

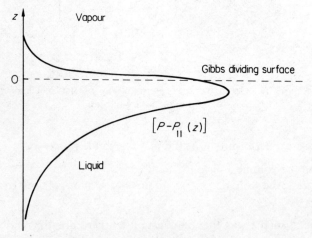

Figure 7.5 Variation of the excess pressure $[P - P_{\parallel}(z)]$
across the free surface

The surface tension can be expressed directly as the integral equation (7.6) together with the normal and tangential components of equation (7.5). Thus, designating the components of the radius vector $\mathbf{r}_{12}$ by (ξ, η, ζ) we have

$$P_{\parallel}(z) = \rho_1(z)kT - \frac{1}{2} \int_0^\infty \rho_{(2)}(\mathbf{r}, z)\nabla\Phi(r)\frac{\xi^2}{r}\,d\mathbf{r} \tag{7.7}$$

$$P = \rho_1(z)kT - \frac{1}{2} \int_0^\infty \rho_{(2)}(\mathbf{r}, z)\nabla\Phi(r)\frac{\zeta^2}{r}\,d\mathbf{r} \tag{7.8}$$

We finally obtain the formal expression for the surface tension

$$\gamma(T) = \frac{1}{2} \int\int_{-\infty}^{\infty} \rho_{(2)}(\mathbf{r}, z)\nabla\Phi(r)\frac{\xi^2 - \zeta^2}{r}\,d\mathbf{r}\,dz \tag{7.9}$$

This expression does not depend upon the location of the dividing surface $z = 0$ relative to the transition zone since $P_\perp = P$ (against which the excess pressure is determined) is uniform along the z-axis. In particular we note that $\gamma(T)$ for classical liquids is not explicitly dependent upon the single-particle density distribution $\rho_{(1)}(z)$. Consequently, numerical agreement between experimental and theoretical estimates of $\gamma(T)$ does not provide a sensitive test of the adequacy of the model transition profiles.

The surface excess energy u_s may be written as the sum of the liquid and vapour components:

$$\begin{aligned} u_s(T) = &\frac{1}{2} \int\int_{-\infty}^{0} [\rho_{(2)}^{L}(\mathbf{r}, z) - \rho_{(2)}^{L}(r)]\Phi(r)\,d\mathbf{r}\,dz \\ &+ \int\int_{0}^{\infty} [\rho_{(2)}^{V}(\mathbf{r}, z) - \rho_{(2)}^{V}(r)]\Phi(r)\,d\mathbf{r}\,dz \end{aligned} \tag{7.10}$$

where $\rho_{(2)}^{L}$ and $\rho_{(2)}^{V}$ represent the liquid and vapour pair density distributions, respectively. It is apparent from equation (7.10) that the surface excess energy is highly sensitive to the location of the origin of coordinates.

Step Model of Liquid Surface

In the absence of an accurate and explicit theory of the functions $\rho_{(1)}(z)$ and $\rho_{(2)}(\mathbf{r}, z)$, Fowler, as an initial approximation, and Kirkwood and Buff, for the purposes of numerical evaluation, resort to the expedient of shrinking the transition zone to a mathematical surface of density discontinuity coincident with the Gibbs dividing surface. It is assumed that the liquid is uniform right up to the boundary, the structure of the transition zone being completely disregarded (Figure 7.6). The conditions on the distribution functions follow as

$$
\left.
\begin{aligned}
\rho_{(1)}(z) &= \rho_L && z < 0 \\
\rho_{(1)}(z) &= 0 && z > 0 \\
\rho_{(2)}(\mathbf{r}, z) &= \rho_{(2)}(r) && z < 0 \quad \text{and} \quad z_P \leqslant 0 \\
\rho_{(2)}(\mathbf{r}, z) &= 0 && z > 0 \quad \text{or} \quad z_P > 0
\end{aligned}
\right\}
\tag{7.11a}
$$

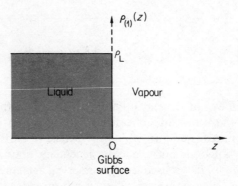

Figure 7.6 Step model of the liquid–vapour density transition. The discontinuity is located on the Gibbs surface

For those regions shown in Figure 7.7(a) where the region of integration is spherically symmetric the contribution to the surface tension in equation (7.9) is zero. However, when $r > z$ as in Figure 7.7(b), going over to spherical coordinates we have

$$
\int_0^{2\pi} \int_{\theta_0}^{\pi} (\xi^2 - \zeta^2) \sin \theta \, d\theta \, d\phi = \pi z (r^2 - z^2)/r
$$

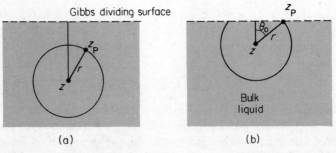

Figure 7.7 Integration geometry for the step model of the liquid surface

so that equation (7.9) becomes

$$\gamma = \frac{\pi}{2}\rho_L^2 \int_{-\infty}^{\infty} z \int_z^{\infty} g_{(2)}^L(r)\nabla\Phi(r)(r^2 - z^2)r\,\mathrm{d}r\,\mathrm{d}z$$

In this approximation Kirkwood and Buff obtain

$$\gamma(T) = \frac{\pi\rho_L^2}{8} \int_0^{\infty} \nabla\Phi(r)g_{(2)}^L(r)r^4\,\mathrm{d}r \qquad (7.11b)$$

for the surface tension at an interface between a liquid phase and a vapour phase of negligible density. For the surface excess energy they obtain

$$u_s = -\frac{\pi\rho_L^2}{2} \int_0^{\infty} g_{(2)}^L(r)\Phi(r)r^3\,\mathrm{d}r \qquad (7.12)$$

These expressions allow relatively simple calculation of the thermodynamic observables γ and u_s from the experimentally determined radial distribution function and the pair potential. The actual transition profile will have a somewhat more relaxed form than that assumed by Kirkwood and Buff, and presumably has a lower free energy. Since the surface tension represents the surface excess free energy per unit area, we can understand that the inequality $\gamma_{(step)} > \gamma_{(expt)}$ is always satisfied. Again, the surface energy will be underestimated in this model, since the excess developed in the vapour phase is entirely neglected, and the inequality $u_{s(step)} < u_{s(expt)}$ is generally satisfied (Table 7.2, SPC entry).

It is found that in the Kirkwood–Buff formulation, agreement between theoretical and experimental values for the surface tension deteriorates with increasing temperature. This is undoubtedly due to inadequate account being taken of the delocalization of the liquid–vapour interface as $T \to T_{crit}$. The Kirkwood–Buff theory retains the density discontinuity throughout: whilst this might be a reasonable approximation at the triple point, it is clearly untenable as the temperature rises.

Linear and exponential profiles have been tried with moderate success, in some cases parametrically adjusted to bring the calculated surface tension and surface energy into agreement with experiment, but these cannot afford much physical insight into the problem.

The theoretical emphasis has shifted from the numerical estimate of the thermodynamic parameters of the surface more toward the determination of the transition profile, to which the thermodynamic parameters are somewhat insensitive.

178

The Single-Particle Distribution $\rho_{(1)}(z)$

The equilibrium spatial delocalization of the liquid–vapour interface develops subject to the thermodynamic constraints of constancy of the chemical potential and normal pressure of the transition zone, and several attempts have been made to solve the non-linear integral equation expressing the constancy of the chemical potential across the transition zone so as to determine the equilibrium density profile $\rho_{(1)}(z)$. The pressure $P(z)$ at any point is then related to the local density through an equation of state, and the surface tension is determined from the pressure-tensor definition (equation (7.6)).

In this quasi-thermodynamic approach the transition zone is imagined to be subdivided into a number of elemental strata, each in thermodynamic equilibrium with its neighbours (Figure 7.8a). Each stratum is characterized

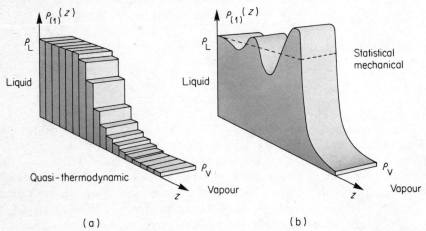

Figure 7.8 (a) The quasi-thermodynamic, (b) the statistical mechanical density transition profile at the free surface of liquid argon

by its local density $\rho_{(1)}(z)$ and pressure $P(z)$, related by the equation of state. The characteristic of the entire transition is, of course, the constancy of the chemical potential. The pressure in each element is not anisotropic however, but is simply related to the local density through the equation of state. In other words, although $P(z)$ will differ from the bulk value P, the condition for mechanical stability of the surface $P_\perp(z) = P$ is not satisfied. Nonetheless, the surface tension is determined as in the pressure-tensor formalism (equation (7.6):

$$\gamma(T) = \int_{-\infty}^{\infty} [P - P(z)]\,dz \tag{7.13}$$

This situation has arisen simply because an equation of state has been used in which the pair distribution is assumed locally isotropic.

In the vicinity of the critical point the transition zone is gradual and extended, and under these circumstances it is legitimate to subdivide the interphasal region into thermodynamic elements over which the density varies inappreciably. However, near the triple point the quasi-thermodynamic approach must be abandoned and a more rigorous statistical mechanical formulation adopted. The dangers in applying macroscopic thermodynamics in regions where molecular density varies by a factor $\sim 10^3$ over a few molecular diameters enforces an approach at a microscopic rather than a macroscopic level. Unfortunately, the two approaches yield quite distinct results, particularly as the triple point is approached. In general, the quasi-thermodynamic analyses tend to yield monotonic profiles (Figures 7.8a, 7.9) whilst the statistical mechanical

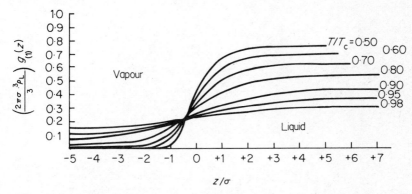

Figure 7.9 Relaxation of the quasi-thermodynamic transition profiles with increasing temperature

approach generally results in an oscillatory transition (Figure 7.8b). Obviously the quasi-thermodynamic approach cannot hope to reproduce the structural features of the transition at a molecular level, and consequently these density profiles are smoother. That is not to say, of course, that these theories are incapable of yielding accurate values of the surface tension and surface energy: these quantities are determined as integrals over the distributions, and to that extent γ and u_s are relatively insensitive to the detailed features of the density transition. It is precisely because of this that the theoretical emphasis has shifted to the determination of $g_{(1)}(z)$.

The statistical mechanical approach consists in solving the single-particle (one-dimensional) Born–Green–Yvon integrodifferential equation which, generalized to the inhomogeneity at the liquid surface, becomes (see equation (3.29))

$$kT\nabla\rho_{(1)}(\mathbf{1}) + \int_2 \nabla\Phi(12)\rho_{(2)}(\mathbf{1,2,12})\,\mathrm{d}\mathbf{2} = 0 \qquad (7.14)$$

This equation is exact, and since it does not contain three-particle distributions there are no difficulties of closure. However, in the absence of an accurate and explicit knowledge of the anisotropic two-particle distribution, we have to assume that $\rho_{(2)}$ retains its bulk form right up to a discontinuity of density coincident with the Gibbs surface, or to form some interpolated structure between the two limiting bulk values, $\rho^L_{(2)}$ and $\rho^V_{(2)}$.

Croxton and Ferrier have retained the isotropic bulk distribution function $g_{(2)}(r_{12})$, but have introduced an angle-dependent coupling operator. Thus, any particle in the vicinity of the plane $z = 0$ is anisotropically coupled to its neighbours, even though the *distribution* of neighbours $g_{(2)}(r_{12})$ remains characteristic of the bulk. This is shown in Figure 7.10 where the coupling varies

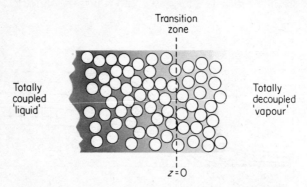

Figure 7.10 Anisotropic molecular coupling at the liquid surface. The density of shading indicates the strength of the coupling varying from the totally coupled situation in the bulk liquid to total decoupling in the vapour

from unity, corresponding to unimpaired (i.e. total) coupling, to zero, in the vapour phase, corresponding to total decoupling. The shading represents schematically the variation of coupling across the plane $z = 0$. Such a device simulates a free surface and all that remains is to determine the analytic variation of the coupling parameter. Clearly there will be three distinct regions to the coupling parameter $\xi(z)$ corresponding to the isotropic bulk liquid phase, $\xi(z) = 1{\cdot}0$, the isotropic bulk vapour phase, $\xi(z) \sim 0^+$ and the transition region, $1 < \xi(z) < 0$ (Figure 7.11). An analytic form similar to, though not necessarily that of, the attractive Lennard–Jones component of the pair potential is assumed for the transition region. This method avoids the reduction of the liquid surface to a step function, although this does represent a special case of the function $\xi(z)$.

The coupling operator is introduced into the BGY equation in the following way

$$\rho_L kT \nabla g_{(1)}(z_1) + \rho_L^2 \int_2 \nabla\{\Phi(r_{12})\xi(z_2)\}g_{(2)}(r_{12})\,\mathrm{d}2 = 0$$

i.e.

$$\frac{kT \nabla g_{(1)}(z_1)}{g_{(1)}(z_1)} + \frac{\rho_L}{2}\int_2 \{\Phi\nabla\xi + \xi\nabla\Phi\}g_{(2)}(r_{12})\,\mathrm{d}2 = 0 \qquad (7.15)$$

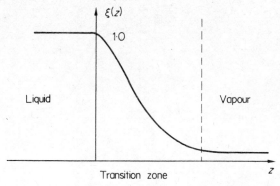

Figure 7.11 Spatial variation of the coupling operator $\xi(z)$ in the Croxton–Ferrier model

where ρ_L is the bulk liquid number density. Integration of equation (7.15) subject to the boundary condition $\rho_{(1)}(-\infty) = \rho_L$, yields

$$\rho_{(1)}(z_1) = \rho_L \exp\left\{ -\frac{\rho_L}{2kT} \int_{-\infty}^{z_1} \int_2 \{\Phi\nabla\xi + \xi\nabla\Phi\} g_{(2)}(r_{12})\, d2\, dz \right\} \quad (7.16)$$

The integral separates into two parts S_1 and S_2 which are attributed to $\Phi\nabla\xi$ representing the constraining effect of the operator at the liquid surface, and $\xi\nabla\Phi$ which represents the modification of the nearest-neighbour force, respectively.

The component S_1 is responsible for the deviation from the simple monotonic form of $g_{(1)}(z)$ (Figure 7.12a): the negative region being responsible for the discrete peak in the density transition profile shown for liquid argon at the triple point. The development of stable density oscillations at the liquid surface must be ascribed directly to the $g_{(2)}(r)\Phi(r)$ product which arises from the $\Phi\nabla\xi$ term in the integrand.

We may understand this layering to develop at the surface as a result of the constraining influence of the bulk attractive field. An assembly of hard spheres for example, constrained by a rigid boundary, develops two or three density oscillations before adopting the uniform isotropic bulk structure (Figure 7.13). Rather than a rigid boundary, Bernal has observed essentially the same phenomenon at the surface of a large spherical assembly of rigid spheres constrained within a balloon, this representing the effect of an internal attractive field and the consequent minimization of surface free energy, so the structure even develops subject to a 'soft' constraint. Evidently the details of the attractive field at the free surface govern the development of the surface structure, subject of course to thermal disruption, and clearly systems capable of developing strong constraining fields, i.e. those having large $\nabla\Phi(r)_{\text{attractive}}$, are the most likely to develop the oscillatory structure, and these primarily are the liquid metals. Such strongly oscillatory profiles as shown in Figure 7.15 would have

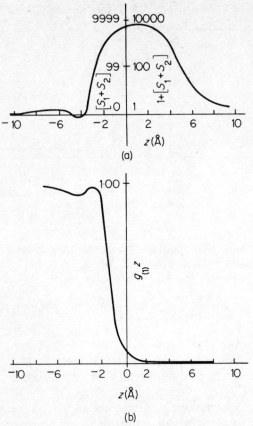

Figure 7.12 (a) Components S_1 and S_2 in the integrand of equation (7.16). (b) The Croxton–Ferrier transition profile for liquid argon at the triple point

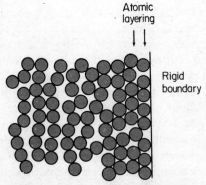

Figure 7.13 Density oscillations in a hard-sphere assembly at a rigid constraining boundary

profound consequences for the thermodynamic functions of liquid metals, as we shall see.

Of course, as the temperature increases so the profile relaxes in accordance with equation (7.16), until at the critical temperature the monotonic variation of density across the transition zone is regained in agreement with the quasi-thermodynamic analyses.

A molecular dynamics simulation of the liquid-argon surface at the triple point has been performed by Croxton and Ferrier, and the stable oscillatory structure confirmed (see Chapter 10). Their computations were for two dimensions only: more recently Barker and Liu have separately performed three-dimensional Monte Carlo and Opitz molecular-dynamic simulations of the liquid-argon surface, and again the stable density oscillations are confirmed (Figure 7.14), although it is somewhat surprising that the structure should extend some ten or eleven atomic diameters into the liquid.

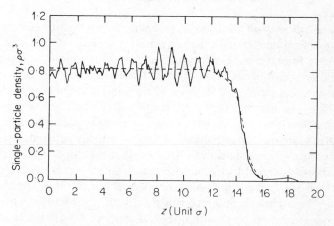

Figure 7.14 Three-dimensional Monte Carlo argon transition profile showing pronounced oscillations (see Figure 10.12)

The Liquid-Metal Surface

It was indicated in the last section that certain liquid systems, notably the liquid metals, might develop strongly oscillatory transition profiles of the schematic form shown in Figure 7.15. This structural development is only likely to occur in the vicinity of the triple point, and then only for systems possessing strong surface fields: at higher temperatures the structure will relax into the familiar monotonic profile. In the case of liquid metals, however, the oscillatory *ionic* distribution $\rho^i_{(1)}(z)$ is likely to be modified by the distribution of conduction electrons at the liquid surface. Since those electrons participating in the conduction process will have energy $\sim E_F$, the Fermi energy, they are effectively free and monochromatic with wavelength $\lambda_F \sim 2\pi/k_F$ where $E_F = \hbar^2 k_F^2/2m$. The diffraction of the plane electronic wavefunctions at a sharp surface boundary

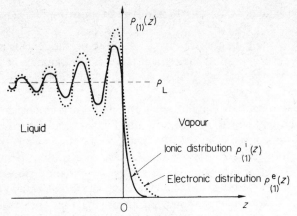

Figure 7.15 Schematic variation of the ionic and electronic profiles at the liquid-metal surface

will establish a density distribution $\rho^e_{(1)}(z)$ analogous to the Fraunhofer intensity distribution of monochromatic light at a straight edge. Quite clearly, when $\lambda_F \sim d$ (mean interatomic spacing in liquid at triple point) a 'resonance' in the distributions can occur and adjust self-consistently so as to enhance the purely geometrical structural features of the ionic distribution at the liquid-metal surface. The maxima in the electronic distribution would preferentially locate on the ionic maxima, and this situation arises in particular when $\lambda_F \sim d$.

The thermodynamic consequences of these relatively ordered surface states are considerable. From equation (7.4) we see that the slope of the $\gamma(T)$ characteristic is directly related to the surface excess entropy per unit area. At temperatures well above the triple point thermal delocalization of the free surface ensures that the surface excess entropy is positive, and the surface tension shows the usual monotonic decrease with increasing temperature going to zero at the critical temperature as $(T_c - T)^\mu$ where $\mu \sim 1.2$.

In the case of the structured transition zone the excess entropy may be negative, in which case a $\gamma(T)$ characteristic of *positive* slope may actually be observed (Figure 7.16). As the structure relaxes with increasing temperature so the $\gamma(T)$ curve passes through an inversion at T_i and eventually regains its 'classical' monotonic decreasing variation with temperature.

From Table 7.1 we see that the metals Zn, Cd and Cu are clearly anomalous in as far as they exhibit an initial positive slope, subsequently inverting to yield the familiar classical behaviour. The excess surface entropy corresponding to the area occupied by a single atom is also given, and the figures clearly reflect the development of low-entropy quasi-crystalline states at the liquid surface. Indeed, the entropy change on melting is $\sim 1.2k$ per particle and the entropy defect at the surface of Zn, Cd and Cu therefore implies a crystalline arrangement of the first two or three atomic layers.

In a series of precision measurements of the surface tension by the sessile drop method, White has demonstrated the importance of thermodynamic

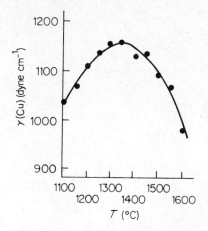

Figure 7.16 Inversion in the surface
tension of copper

Table 7.1. Surface entropies of liquid metals at the
triple point (Faber)

	$d\gamma/dT$ (dyne/cm/°C)	S_s/atom (units of k)
Li	−0·14	0·80
Na	−0·1	0·85
K	−0·06	0·78
Rb	−0·06	0·94
Cs	−0·05	0·78
Ag	−0·13	0·65
Al	−0·135	0·68
In	−0·096	0·61
Sn	−0·083	0·55
Cu	+0·75	−2·9
Zn	+0·5	−2·2
Cd	+0·5	−2·9

equilibrium of the liquid surface with its vapour. The surface tension is, of course, an equilibrium thermodynamic parameter, and yet most experimental arrangements neglect the precaution of ensuring that there is no net loss of vapour across the free surface.

The effect of continuous vaporization on the surface tension of zinc is shown in Figure 7.17(a). White's method was to enclose the sessile drop in an isothermal quartz cell and photograph its profile (Figure 7.17b). Deliberately non-equilibrium measurements could be made by sliding back the lid of the cell leading to a net transport of vapour.

The effects are readily understood by considering the schematic entropy-temperature curves. In Figure 7.18(a) we show the 'classical' variation of the

186

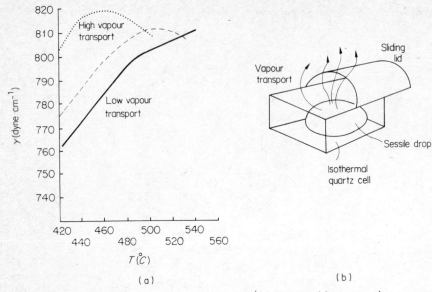

Figure 7.17 (a) Variation of the surface tension of zinc with progressive vapour transport. At high net transport rates the $\gamma(T)$ characteristic shows the 'classical' monotonic decrease with temperature. (b) Quartz isothermal cell containing sessile liquid metal droplet

excess entropy with temperature as the difference in the bulk (S_β) and surface entropies (S_σ). Here $S_\sigma > S_\beta$ over the entire temperature range, and from equation (7.4) the slope $d\gamma/dT$ is evidently a monotonic negative function of temperature, vanishing at the critical point.

In the inverted case $S_\sigma < S_\beta$ just above the triple point, although the rate of entropy production at the free surface will be greater than that of the bulk, and consequently the bulk and surface entropy curves, initially inverted, cross and reinvert to yield classical behaviour.

Now, consider an inverting system such as zinc, but now allow a net transport of vapour across the liquid surface—a degree of non-equilibrium. This will not affect the bulk states of course, but it will result in the upward bodily shift of the surface entropy curve from S_{σ_1} to S_{σ_2} corresponding to the 'randomization' of the free surface. The effect will be a *lowering* of the inversion temperature, and under sufficiently non-equilibrium circumstances, the inversion may disappear altogether, the system then exhibiting classical behaviour. This is seen to account for the curves shown in Figure 7.17(a), and raises some doubt as to the adequacy of the general body of surface-tension data.

One other striking qualitative observation concerns the form of the sessile drop on solidification. Those systems showing pronounced positive $\gamma(T)$ slopes solidify into beautifully faceted droplets (Figure 7.19) unlike those showing classical behaviour which solidify into perfectly smooth drops. It is difficult to escape the conclusion that the development of stable density oscillations at

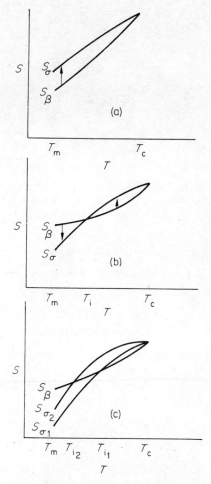

Figure 7.18 (a) 'Classical' behaviour of the bulk and surface entropies as a function of temperature. The difference $S_\beta - S_\sigma = \partial\gamma/\partial T$ shows a monotonic decrease in γ with increasing T. (b) $S_\sigma < S_\beta$ inversion at T_m. Differing rates of entropy production produce inversion at T_i. $\gamma(T)$ is almost linear. (c) Increasing rate of vapour transport shifts the S_σ curve up the entropy axis, and T_i to lower values as observed by White for zinc (Figure 7.17a)

the liquid surface serves as a nucleation centre in the solidification process. Indeed, on sectioning a zinc droplet, it is found that the growth of the single crystals initiated at the liquid surface. X-ray reflection investigations show that the facets are basal planes—planes of maximum packing density.

Figure 7.19 'Geodesic' faceting at the surface of a solidified sessile drop of liquid zinc. Each facet is a basal plane

The Two-Particle Distribution $\rho_{(2)}(z, \mathbf{r})$

Before a complete statistical mechanical calculation of the surface tension can be concluded on the basis of equation (7.9) we require the anisotropic two-particle distribution

$$\rho_{(2)}(z_1, \mathbf{r}_{12}) = \rho_{(1)}(z_1)\rho_{(1)}(z_2)g_{(2)}(z_1, \mathbf{r}_{12}) \tag{7.17}$$

Assuming we know the single-particle distribution $\rho_{(1)}(z)$ the problem reduces to the determination of $g_{(2)}(z_1, \mathbf{r}_{12})$. The usual hierachical relation holds between adjacent orders of distribution (equation (3.4)) and we may test our prescription for $g_{(2)}$ against the condition

$$-1 = \int \rho_{(1)}(2)[g_{(2)}(\mathbf{1}, \mathbf{2}) - 1]\,\mathrm{d}\mathbf{2} \tag{7.18}$$

which it should satisfy. None of them do, however, and to that extent the one- and two-particle approximations are inconsistent.

Kirkwood and Buff approximated the anisotropic pair distribution by means of a step model of the liquid surface, equation (7.11b). A realistic representation first proposed by Green is simply to write

$$\rho_{(2)}(z_1, \mathbf{r}_{12}) = \rho_{(1)}(z_1)\rho_{(1)}(z_2)g_{(2)}^{L}(r_{12}) \tag{7.19}$$

where $g_{(2)}^{L}$ is the bulk liquid distribution.

Croxton and Ferrier have expressed the anisotropic pair distribution in terms of the angular distribution

$$\rho_{(2)}(z_1, r_{12}, \theta, \phi) = \rho_L^2 g_{(2)}^{L}(r) \sum_{l}^{\infty} \sum_{m=0}^{l} A_{lm}(z_1, T)P_l^m(\cos\theta)\Phi(m\phi) \tag{7.20}$$

It is assumed that there is no radial distortion of the bulk distribution function $g_{(2)}^{L}(r)$, and the spatially dependent set of coefficients describe how the distortions of the spherically symmetric bulk distribution develops in the vicinity of the

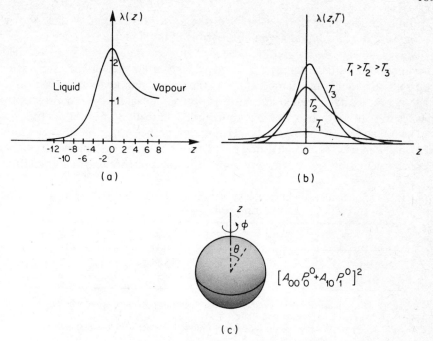

Figure 7.20 (a) The hybridization ratio $\lambda(z)$ for the liquid-argon surface at the triple point. (b) Schematic relaxation of $\lambda(z)$ with increasing temperature as the structural anisotropy relaxes. (c) The sp_z hybrid distribution

transition. Equation (7.19) represents no more than an approximate harmonic mode analysis of the exact distribution, equation (7.17). The distribution is obviously symmetric about the z-axis (for non-hydrodynamic equilibrium systems of pairwise interacting spherical molecules at a planar interface), and therefore $m = 0$ throughout. Again, on symmetry grounds many of the $\{A_{l0}\} = 0$, and Croxton and Ferrier retain only the $l = 0$, $l = 1$ harmonics in their angular description of the distribution in the transition zone (Figure 7.20c):

$$\rho_{(2)}(z_1, \mathbf{r}) = \rho_L^2 g_{(2)}(r)\{A_{00}(z_1)P_0^0(\cos\theta) + A_{10}(z_2)P_1^0(\cos\theta)\}^2$$

$$= \rho_L^2 g_{(2)}(r)A_{00}^2(z_1)\{P_0^0 + \lambda(z_1)P_1^0\}^2 \qquad (7.21)$$

where $P_0^0 = 1$, $P_1^0 = \cos\theta$. The trial angular function has been squared to increase its flexibility a little, and has been rewritten in terms of $\lambda(z_1) = A_{10}(z_1)/A_{00}(z_1)$, which has the significance of a *hybridization coefficient* between the spherically symmetric bulk modes and the surface angular modes of the appropriate symmetry. It is clear that $\lambda \to 0$ as $z \to \pm\infty$: hybridization of bulk liquid and vapour modes with the specifically interphasal configurations will diminish as the bulk and surface correlations decouple far from the interphase, the pair distribution becoming spherically symmetric. The first unassociated harmonic P_1^0 shows extensive hybridization with the spherically symmetric

P_0^0 (bulk) modes only in the vicinity of the anisotropic liquid surface shown in Figure 7.20(a) for liquid argon at the triple point (84 K).

The coefficients $A_{00}(z)$, $A_{10}(z)$ and hence $\lambda(z)$, are determined so as to yield the minimum surface energy—a standard variational technique. $\lambda(z)$ is, of course, also a function of temperature: at the critical temperature the transition zone is broad and extended and little anisotropy develops. $\lambda(z)$ consequently relaxes to $\sim 0^+$ with temperature, shown schematically in Figure 7.20(b).

The full statistical mechanical determination of the surface tension may now be concluded with the results shown in Table 7.2. In the case of the Croxton–Ferrier calculation the anisotropy implicit in the pressure-tensor formulation

Table 7.2. Comparison of results for surface properties of liquid argon at the triple point (84 K)

	H	SPC	KB	CF	Expt
γ (dyn cm^{-1})	21·6	15·6	16·84	13·48	13·45
u_s (erg cm^{-2})	60·59	27·08	44·3	35·35	35·01

H: Hill's quasi-thermodynamic (constant chemical potential) approach. Results corrected by Plesner and Platz.
SPC: Shoemaker *et al.* using Kirkwood–Buff step model with the best available $g_{(2)}(r)$ and $\Phi(r)$ data (1970).
KB: Estimated from Kirkwood and Buff's 90 K values by $\gamma = \gamma_0 (1 - T/T_c)^{1·28}$, $u_s = \gamma - T(\partial\gamma/\partial T)$.
CF: Croxton and Ferrier.

is re-expressed in terms of $\lambda(z)$ which is, of course, a measure of the structural anisotropy. Whilst the agreement appears quite good, it has to be emphasized that the principal thermodynamic quantities are given as *integrals* over the one- and two-particle distributions, and to that extent are relatively insensitive to the structural analysis.

References

C. A. Croxton, *LSP.*
C. A. Croxton, *Adv. Phys.*, **22**, 385 (1973).

CHAPTER 8
Liquid Metals

Introduction

The role of the conduction electrons in establishing the structural and electrical transport properties of liquid metals and binary alloys is by no means fully understood. To a certain extent the difficulty arises from what we already know of the electronic processes in solid crystalline metals. For although in the relatively disordered liquid structure it seems that characteristics of the solid phase such as anisotropic Fermi surfaces and Brillouin zones—properties generally associated with crystal symmetry—will not survive the melting transition, short-range order certainly does exist. Experimental evidence on the basis of the Hall effect and certain optical properties seem to indicate a nearly-free-electron (NFE)-like behaviour, and a plane-wave representation of the conduction electron states rather than Bloch functions would appear to be justified. But how are we to reconcile this with the organization of ionic scatterers in the liquid which exhibit a short-range if not long-range order?

We have already seen from the simple relationship between the experimentally observable direct correlation and the pair potential that the ion–ion interaction is of a fundamentally different form to that operating in an insulating fluid. The long-range form of the interaction appears to be characterized by radially damped *oscillations* whose amplitude seems to decrease with increasing temperature. The amplitude of the oscillations is significantly less than kT, and is unlikely to modify the structural and dynamical features of the assembly—a conclusion reached by Schiff on the basis of molecular dynamic simulation techniques (Chapter 10).

Since the conduction electron states are necessarily orthogonalized to those of the core, a great simplification arises in that the long-range Coulombic (screened-ion) interaction of the conduction electrons with the core is supplemented by a strongly repulsive exclusion from the core region. The result is a weak *pseudopotential* which allows us to work in the Born approximation and employ all the weak scattering perturbative techniques and approximations. We shall have to consider the nature of the scattering potential in some detail since it will prove of central importance in our discussion of electrical resistivity, though we shall find that models advanced on a physical basis prove just as acceptable as the more direct *a priori* calculations.

In this chapter we shall consider only the resistivity problem in any depth since it exposes the main features of the electronic conduction process in disordered structures. Other problems of interest such as thermopower and optical properties should then present no formal difficulty.

The Nearly-Free-Electron Model

What experimental support is there for the assertion that the conduction electron states are nearly-free-electron-like? Perhaps the most direct and convincing evidence is to be had on the basis of Hall-effect measurements. The ratio of the experimental Hall coefficient R_{expt} and the elementary free-electron value $R_{calc} = -1/nec$ is shown for a variety of liquid-metal systems in Table 8.1.

Table 8.1. Experimental and calculated Hall coefficients

R_{exp}/R_{calc}	Liquid	Solid
Li		1·3
Na	0·98	1·04
K		0·90
Cs		1·06
Cu	1·0	0·74
Ag	1·0	0·81
Au	1·0	1·46
Be		−9·80
Zn	1·01	−0·71
Cd	0·99	−0·92
Hg	0·99	1·2
Ga	0·97	
In	0·93	0·45
Tl	0·96	
Ge	1·06	~3 × 10³
Sn	1·0	0·33
Pb	0·88	~0
Sb	0·92	−25
Bi	0·95	−250·00

The ratios are remarkably close to unity in all but one or two liquid-metal systems—and in no case are the anomalous ratios of the solid state observed. Admittedly there are experimental difficulties associated with the measurement of the Hall coefficient—the Lorentz force generates magneto-hydrodynamic circulating currents in the liquid metal—nevertheless the errors are unlikely to exceed 5 per cent. On this basis then there appears reasonable justification in assuming a nearly-free-electron model of electrical transport.

In all but the lightest elements the motion of the ion may be neglected and we restrict the discussion to the *elastic* scattering of an electron from an initial momentum state **k** to a final state **p** in the weakly scattering *total* (collective) pseudopotential, $U(\mathbf{r})$. The matrix element responsible for scattering over the constant-energy Fermi surface (Figure 8.1) in the Born approximation is

$$\langle \mathbf{p}|U|\mathbf{k}\rangle = \frac{1}{\Omega}\int_\Omega \exp(-i\mathbf{K}\cdot\mathbf{r})U(\mathbf{r})\,d\mathbf{r}$$

$$= \frac{U(\mathbf{K})}{\Omega}, \quad \text{where } \mathbf{K} = \mathbf{p} - \mathbf{k}$$

(8.1)

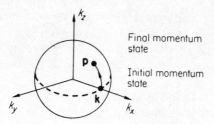

Figure 8.1 Constant energy scattering of an electron from state **k** to state **p** over the isotropic Fermi surface in the field of the pseudopotential

The mean lifetime τ of an electron in state **k** is given by the Fermi 'golden rule'

$$\frac{1}{\tau_{\text{NFE}}} = \frac{mk_{\text{F}}}{2\pi\hbar^3\Omega} \int_0^\pi \overline{|U(K)|^2} \sin\theta \, d\theta \tag{8.2}$$

where we assume $\overline{|U(K)|^2}$ is independent of the direction of K in an isotropic liquid.

In fact, solution of the Boltzmann transport equations shows that we have to weight the integrand appearing in equation (8.2) by the factor $(1 - \cos\theta)$

$$\frac{1}{\tau_{\text{NFE}}} = \frac{mk_{\text{F}}}{2\pi\hbar^3\Omega} \int_0^\pi \overline{|U(K)|^2} \sin\theta(1 - \cos\theta) \, d\theta$$

$$= \frac{mk_{\text{F}}}{\pi\hbar^3\Omega} \int_0^1 \overline{|U(K)|^2} \, 4\left(\frac{K}{2k_{\text{F}}}\right)^3 \cdot d\left(\frac{K}{2k_{\text{F}}}\right) \tag{8.3}$$

and the resistivity ρ then follows from the elementary free-electron formula

$$\rho = \frac{m}{ne^2\tau} \tag{8.4}$$

where τ is now taken to be τ_{NFE}.

It remains to specify $U(\mathbf{r})$ representing the collective pseudopotential and implicitly containing the details of the ionic configuration in the liquid. $U(\mathbf{r})$ is then the sum of the individual (local) screened psueodopotentials $u(\mathbf{r} - \mathbf{R}_l)$, where $\mathbf{R}_l$ denotes the configuration of the N ions in the volume Ω. The Fourier transform follows quite simply as the ensemble average over the various ionic configurations:

$$\overline{|U(K)|^2} = N|u(K)|^2S(K) \tag{8.5}$$

where $S(K)$ is the ionic structure factor for the system and $u(K)$ is the psueodopotential due to a single ion. We therefore finally arrive at the Ziman formula for the resistivity:

$$\rho = \frac{m}{ne^2\tau_{\text{NFE}}} = \frac{m^2N}{ne^2} \frac{k_{\text{F}}}{\pi\hbar^3\Omega} \int_0^1 |u(K)|^2S(K)4\left(\frac{K}{2k_{\text{F}}}\right)^3 d\left(\frac{K}{2k_{\text{F}}}\right) \tag{8.6}$$

194

If the distribution of scatterers were purely random ($S(K) = 1.00$) then (8.6) would yield the familiar result that the total scattered wave intensity is just the sum of the individual scattered intensities. More generally, the configurationally dependent interference is expressed in terms of the interference function or structure factor, $S(K)$.

It now only remains to specify the pseudopotential $u(K)$, and the liquid-metal resistivity follows from the numerical evaluation of the Ziman formula, equation (8.6).

Choice of the Scattering Potential $u(K)$

The detailed choice of $u(K)$ is a problem in pseudopotential theory, but as a first approximation we might suggest a screened Coulomb point-ion form

$$u(r) = -\frac{Ze^2}{r} \exp(-q_s r) \tag{8.7}$$

which is, of course, the familiar Debye–Hückel result (Chapter 2). Z is the valency of the liquid metal and q_s the screening length. For the Fourier transform $u(K)$ we find from (8.7)

$$u_{\text{point ion}}(K) = -\frac{N}{\Omega} \frac{4\pi Z e^2}{K^2 + q_s^2} \tag{8.8}$$

and we show this function in Figure 8.2. Since NZ is the total number of con-

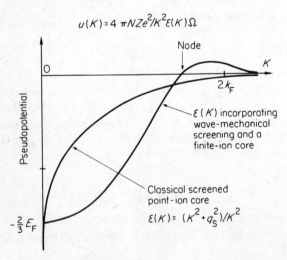

Figure 8.2 The pseudopotential $u(k)$ for a classically screened point-ion core, and for a wave-mechanically screened finite-ion core. The latter model induces a node at $K < 2k_F$, and a subsequent positive region

duction electrons we have

$$u_{\text{point ion}}(K) = -\frac{4k_F^3 e^2}{3\pi(K^2 + q_s^2)} \qquad (8.9)$$

and in the limit $K \to 0$

$$u_{\text{point ion}}(0) = -\tfrac{2}{3}E_F \qquad (8.10)$$

Whatever the details of the pseudopotential in the vicinity of the core, at large distances $u(r)$ will behave like a real screened Coulomb potential round a point charge Ze, and therefore all models will yield the limiting result (8.10) as $K \to 0$. At large K the details of the core features will be apparent.

Equation (8.9) may be conveniently expressed in terms of a general dielectric function $\varepsilon(K)$, i.e.

$$u(K) = -\frac{4\pi Ze^2 N}{\Omega K^2 \varepsilon(K)} \qquad (8.11)$$

where $\varepsilon(K) = (K^2 + q_s^2)/K^2$ represents the Fourier transform of the classical dielectric function, and more generally expresses the departure from the point-ion model.

We need to incorporate the effects of a *finite core* before we are in a position to relate our discussion to the calculation of liquid-metal resistivity. In particular, as we anticipated above, this will lead to a modification in the large-K region of the classically screened point-ion pseudopotential, whilst in the long-wavelength limit $(K \to 0)$ $u(K)$ will remain essentially unchanged $(u(0) = -2E_F/3)$.

The effect of introducing a finite real-space core is to establish a damped oscillation in $u(K)$, or in particular a *node* followed by a positive region (Figure 8.2) which, as we shall see, proves crucial in our calculation of electrical resistivity. The location of the node, and the detailed form of $u(K)$ are, however, still under discussion.

Calculation of the Resistivity

At low temperatures the resistivity of the solid is generally governed by residual scattering of the conduction electrons by crystal imperfections, dislocation and impurities. With increasing temperature electron–phonon scattering processes attributed to the thermal lattice vibrations also contribute: both components represent 'disorder scattering' which would be expected to increase on melting, and indeed the resistivity generally doubles (approximately) as shown in Figure 8.3. There are some notable exceptions, the semi-metals Sb, Bi and Ga, for example, but these crystallize into abnormally open solid structures, and the coordination actually *increases* on melting to form a 'normal' liquid.

The temperature coefficient of resistivity $(\partial\rho/\partial T)_P$ would, on the above model, be expected to be positive in both phases, although here again, Zn and Cd are exceptional in that they yield initially negative coefficients just beyond the melting point.

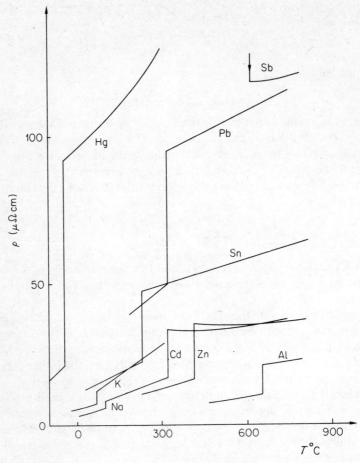

Figure 8.3 Variation of resistivity with temperature. All systems show a discontinuity in the resistivity on melting—generally an increase, although some semi-metals with anomalous solid structures show a *decrease* on melting. Nearly all show positive temperature coefficients, except Zn and Cd

This is the background of experimental data against which we have to test the Ziman formula, and we shall find that the agreement is good, even showing the correct qualitative dependence upon valency and the 'anomalous' temperature coefficients of Cd and Zn.

For the evaluation of the Ziman integral (8.6) we require a knowledge of the structure factor $S(K)$ and the pseudopotential $u(K)$. The former varies little from system to system at the melting point, but the upper limit of the integration, $2k_F$, depends sensitively on the *valency* of the metal and, of course, the sharpness of the Fermi surface. Since the Fermi radius k_F varies with valency as $Z^{\frac{1}{3}}$, the location of the cut-off depends on whether the liquid is mono-, di- or polyvalent (see Figure 8.4a). On the basis of a $u(K)$ of the form shown in Figure 8.2, the

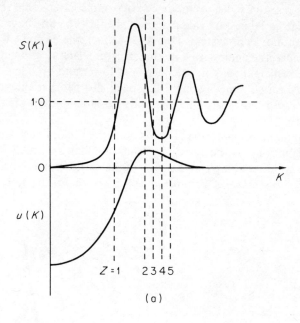

$S(K)$

1·0

0

$u(K)$

$Z = 1$ 2 3 4 5

(a)

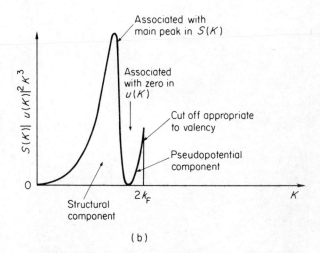

Associated with
main peak in $S(K)$

Associated
with zero in
$u(K)$

Cut off appropriate
to valency

Pseudopotential
component

$S(K)|u(K)|^2 K^3$

0

$2k_F$

K

Structural
component

(b)

Figure 8.4 (a) Typical structure factor $S(K)$ and pseudopotential $u(K)$. The location of $2k_F$ for metals of valency $Z = 1, 2, \ldots$ is shown by the broken lines. (b) The product $S(K)|u(K)|^2 K^3$ appearing in the Ziman integrand, with the cut-off appropriate to the valency. A qualitative resolution of the integrand into a long-wavelength structurally dependent component, and an intermediate-wavelength component sensitive to the core details of the pseudopotential

Ziman integrand is shown in Figure 8.3(b): the area beneath the curve is proportional to the resistivity of the liquid. We are immediately able to explain, for example, the fact that polyvalent liquid metals (Pb, Sn) have a higher resistance than monovalent systems (Na, Ag)—in fact, the cut-off value $2k_F$ in silver occurs *before* the main peak in the structure factor. With increasing temperature the principal peak in $S(K)$ becomes lower and broader, and since the zinc cut-off is located centrally on the first peak, the theory is able to account for the initial *decrease* in the resistivity of liquid zinc with increasing temperature (Figure 8.3).

We are able to make a qualitative subdivision of the Ziman integrand (Figure 8.4b) into contributions to the resistivity from small-angle (small-K) *plasma scattering* in the collective long-range region of the pseudopotential, and large-angle (large-K) *core scattering*. From Figure 8.5 we see, for example, that the

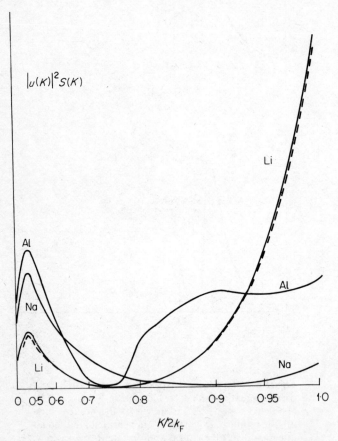

Figure 8.5 Variation of $|u(K)|^2 S(K)$ with $K/2k_F$ for Al, Na and Li. The distribution of area beneath the curves indicates whether the resistivity is due primarily to large- or small-angle (K) scattering. The broken curve for Li shows the effect of correcting for ionic recoil in the scattering process

resistivity of Na is dominated by plasma scattering whilst in Li core-scattering processes dominate. The broken curve shown for Li shows the effect of correcting for ionic recoil in the scattering process: obviously this will be significance for the lightest metallic systems.

Of course, the value of the resistivity calculated on the basis of the Ziman formula depends sensitively on the scattering data and the choice of pseudo-potential. Nevertheless, the agreement between theory and experiment is quite reasonable, as we see from Table 8.2, and certainly much more satisfactory than the corresponding calculations of the solid resistivity.

Table 8.2. Resistivities at the melting point ($\mu\Omega$ cm) (after Faber)

	ρ (expt)	ρ (calc)	Valency
Li	24·7	25	1
Na	9·6	7·9	1
K	13·0	23	1
Rb	22·5	10	1
Cs	36·0	10	1
Zn	37	37	2
Cd	34	23	2
Hg (25 °C)	91	30	2
Al	24	27	3
In	33	24	3
Tl	73	60	3
Pb	95	64, 77	4
Bi	128	87	5

As we pointed out, one of the central assumptions made in the application of the Ziman theory is that of a *sharp* Fermi surface; however, there is some evidence of *blurring* of the Fermi surface as evidenced in positron annihilation experiments, and this we shall review in some detail below.

More About $u(K)$

As soon as we introduce a *finite* core with a discontinuous (or at least sharp) boundary, oscillations are induced in $U(\mathbf{K})$ (Figure 8.2). The resulting node and positive region prove crucial in the evaluation of Ziman's integral for the liquid resistivity, and serves to *lower* ρ significantly below its classical estimate. As we see from Figure 8.4(b), the integrand separates quite clearly into two distinct regions originally designated the *plasma* and the *pseudopotential* components. The small-K part obviously arises mainly from the screened Coulombic region of the potential and the structure of the liquid, whilst the large-K component arises from the effectively incoherent scattering from the cores.

The original work of Ziman explicitly resolved $U(K)$ into two components

$$U(K) = U_1(K) + U_2(K) \qquad (8.12)$$
$$\text{plasma} \quad \text{core}$$
$$\text{term} \quad \text{term}$$

200

and whilst this provided a useful physical resolution of two of the main processes governing the resistivity, the distinction is of a purely qualitative nature and has since been replaced by more subtle approaches which do not depend upon an arbitrary subdivision into plasma and core processes.

Direct calculations of the pseudopotential have been attempted in which the true electronic wavefunction ψ is decomposed into a long-range asymptotic component, ϕ, which varies smoothly in the core region, and a residue in the core region which is expressed in terms of the core functions ψ_c:

$$\psi = \phi - \sum_c a_c \psi_c \qquad (8.13)$$

ψ is orthogonal to all the core functions, and the core functions are mutually orthogonal. Multiplying (8.13) throughout by ψ_c and integrating therefore gives

$$0 = \int_\Omega \phi \psi_c \, d\tau - \sum_{c'} a_{c'} \int \psi_c \psi_{c'} \, d\tau$$

so that

$$a_c = \int_\Omega \psi_c \phi \, d\tau \qquad (8.14)$$

Substitution of (8.13) into Schrödinger's equation $H\psi = E\psi$ yields, after some simple manipulation,

$$\left\{ H + \sum_c \frac{(E - E_c)a_c\psi_c}{\phi} \right\} \phi = E\phi \qquad (8.15)$$

where a_c is given by (8.14). The expression in the curly brackets is a *pseudo-Hamiltonian* and contains the extra positive term which plays the role of an exclusion potential, leaving only a relatively weak pseudopotential from the conduction electron's point of view.

If we wish to calculate the pseudopotential explicitly we have, of course, to determine the second term in the pseudo-Hamiltonian. We might propose that the asymptotic pseudowavefunction ϕ is reasonably represented by the plane wave $\exp(i\mathbf{k} \cdot \mathbf{r})$, and knowing the core functions ψ_c we have a route to the pseudopotential operating in the liquid metal. There are numerous possible variations and approximations on this approach, all differing slightly in the final estimate of the pseudopotential, and leading to resistivities differing by a factor of two or three.

Certainly more convenient, and just as reliable, are the model pseudopotentials advanced on largely physical grounds—the parameters of the potential being adjusted to yield the correct band gaps in the electronic energy spectrum of the solid, or so as to fit the observed solid phonon dispersion curves or even the correct liquid resistivity! Examples of such potentials are shown in Figure 8.6. Experimentally the blurring may be investigated by positron annihilation. A positron of several MeV is fired into a liquid metal when, after a number of inelastic encounters, it slows down and finally annihilates with a core or

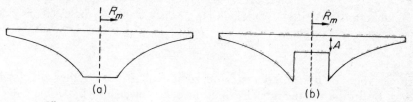

Figure 8.6 Model real-space pseudopotentials (a) Shaw, (b) Heine

conduction electron, emitting two or three photons in a momentum-conserving process. The 'terminal' velocity of the positron is likely to be very small relative to that of the electron and for the two-photon mode of decay, the small angle θ (Figure 8.7a) indicates that the pair (e^+, e^-) had linear momentum $\hbar k$ just prior

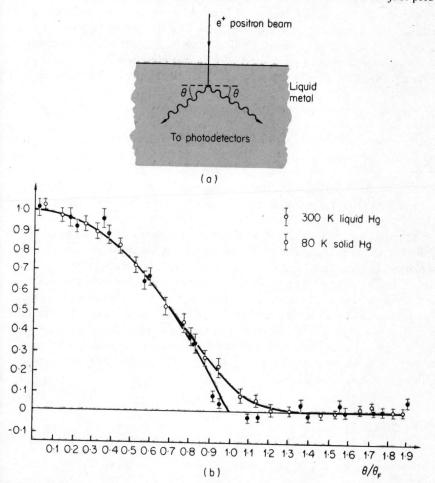

Figure 8.7 (a) Two-photon mode of decay of the electron–positron annihilation. (b) Angular distribution of photons for solid and liquid mercury. The tail is attributed to a blurred Fermi surface

to annihilation, virtually all of which may be attributed to the electron. Since linear momentum is conserved

$$\hbar \mathbf{k} = 2mc \sin \theta \sim 2mc\theta \tag{8.16}$$

The variation of the photodetector current with θ should then give some indication of the momentum distribution of the electrons. Experimentally it appears that the blurring of the Fermi surface in Hg is ~ 20 per cent, and the effect on the ion–ion interaction is to suppress the Freidel oscillations altogether (Figure 8.8). The stability of liquid mercury then presents some difficulty, although the delocalization of the valence electrons throughout the volume of the metal confers a degree of stability. Nevertheless, the interpretation of the experimental data is open to some reconsideration. For example, whilst the positron is electrostatically excluded from the core, some contribution must certainly be attributed to the core electrons, and the two components cannot be isolated. Again, the positron itself may seriously perturb the momentum distribution, and, indeed, may modify the ionic structure of the liquid in its immediate vicinity, all of which may modify the structure of the Fermi surface.

Alloys

The resistivities of *solid* binary alloys are found to obey two rules relating to the atomic concentration, c. The first, due to Nordheim, gives the dependence of the resistivity ρ on c as $c(1 - c)$. A second rule, due to Linde, relates the valence difference between solvent and solute to $d\rho/dc$, this only for dilute alloys.

Experimentally it is found that these rules are only obeyed for *liquid* systems when the solvent is monovalent.

The resistivity of a pure liquid metal in Ziman's treatment is given by equation (8.6). It would appear then that for a binary alloy AB we shall need two pseudo-potentials $U_A(K)$, $U_B(K)$ and three partial-structure factors $S_{AA}(K)$, $S_{AB}(K)$ and $S_{BB}(K)$. Only by the application of three distinct diffraction techniques (X-rays, neutrons and electrons, say) may the three partial-structure factors be isolated 'simultaneously'. Whilst such partial factors exist, a qualitative understanding of the liquid-alloy behaviour has been given by Faber and Ziman. If it is assumed that the two atomic constituents A and B have the same atomic volume and valency, then we may set $S_{AA} = S_{AB} = S_{BB} = S$. The resistivity may then be written

$$\rho = \rho_1 + \rho_2 \tag{8.17}$$

where

$$\rho_1 = [(3\pi/\hbar e^2)/\rho_0 v_F^2)] \langle (1 - c)SU_A^2 + cSU_B^2 \rangle \tag{8.18}$$

$$\rho_2 = [(3\pi/\hbar e^2)/(\rho_0 v_F^2)] \langle c(1 - c)(1 - S)(U_A - U_B)^2 \rangle \tag{8.19}$$

The second term is seen at once to offer an explanation for the Nordheim dependence on concentration, whilst the first term describes a linear interpolation between ρ_A and ρ_B: which mode of behaviour is in fact observed depends on the relative magnitudes of ρ_1 and ρ_2.

Now while $S(K)$ is not small, we see from Figure 8.4(a) that for monovalent metals it is less than unity over most of the range $0 < K < 2k_F$. It therefore follows that although ρ_2 does not necessarily dominate ρ_1, it may nevertheless ensure that the resistivity isotherm (8.17) has a convex form (Figure 8.9a). If

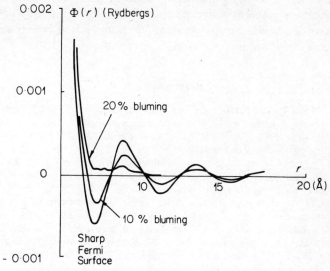

Figure 8.8   The effect of progressive Fermi surface blurring on the Friedel oscillations in liquid mercury

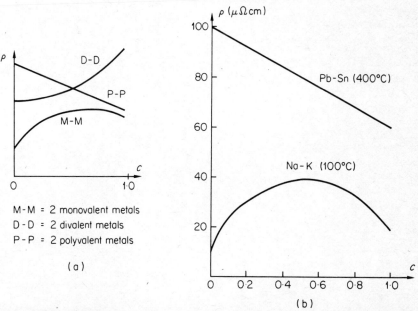

Figure 8.9   (a) Qualitative variation of binary alloy (AB) resistivity with concentration c. (b) Examples of a binary monovalent (Na–K) and binary polyvalent (Pb–Sn) system

204

the two metals are divalent, the limit in the resistivity integral, $2k_F$, lies in a region for which $S(K) > 1$. Bearing in mind that the integral is heavily weighted in this region, ρ_2 could become negative yielding the *concave* isotherm shown in Figure 8.8(a). For systems of higher valency $S(K) \sim 1.0$ over most of the range of K that matters, and ρ_2 should play a subordinate role. These qualitative predictions are in agreement with experiment as we show in Figure 8.9(b), and form the basis of a qualitative understanding of associated phenomena such as thermopower in liquid binary alloys.

References

N. H. March, *Liquid Metals*, Pergamon (1968) provides a useful and readable introduction to the structure, dynamics and electrical properties of liquid metals.

N. Cusack, *Rep. Prog. Phys.*, **26**, 361 (1963). A review of the theoretical and experimental status of the physics of liquid metals.

T. E. Faber, *Theory of Liquid Metals*, Cambridge University Press (1972). An advanced text providing an extensive review of the theoretical and experimental development of the physics of liquid metals and alloys.

C. A. Croxton, *Introductory Eigenphysics*, Wiley (1974). Gives an elementary partial wave analysis of the development of Friedel oscillations in the context of general scattering theory. The nature and properties of the Fermi surface are also discussed.

T. E. Faber, in *The Physics of Metals* (Ed. J. M. Ziman), Cambridge University Press (1969), pp. 282–316. A review at intermediate level of the current theoretical status of electrical transport properties in liquid metals.

CHAPTER 9

Liquid Crystals

Introduction

The liquid-crystal systems are characterized by two or more distinct melting points forming intermediate *mesophases* between the highly ordered crystalline solid and the isotropic liquid. Cholesteryl benzoate, for example, melts at 145 °C to form a turbid, optically anisotropic liquid which, on heating to 179 °C undergoes a second transition to yield a second isotropic clear fluid. Optical birefrigence is more characteristic of crystalline solids and yet these organic compounds exhibit many features generally taken to characterize the liquid phase.

The molecular requirement for the existence of a mesophase is a pronounced rod-like structure, and the liquid crystalline properties arise from the strong tendency for the molecules to lie with their long axes aligned. So, although spatial order of the crystalline phase is destroyed on melting, *angular* correlation survives and yields a mesophase between the solid and isotropic liquid phases which persists until thermally disrupted into the isotropic phase. The stability and thermal range of the mesophase is governed largely by the geometric features of the molecule, as we shall see from the computer simulations of Chapter 10.

There are a number of liquid-crystal morphologies which can arise, and these were broadly classified by Friedel in 1922. The simplest is the *nematic* mesophase which forms a turbid, low-viscosity liquid. At a molecular level the nematics are characterized by a disordered distribution of molecular centres with pronounced molecular alignment. The *cholesteric* phase, formed by many esters of cholesterol which in itself is not a liquid crystal, is often considered as a twisted nematic: the angular orientation shows a helical spatial dependence which, under the effect of electric or magnetic fields, may be stretched out to form a nematogen. The final major class of mesogens is the *smectic* phase which is qualitatively subdivided into subphases A to H, although only the structures of the first few phases (A to D) are known with any certainty. The smectics are most akin to the solid phase, certain smectogens showing extended three-dimensional order, both in the distribution of molecular centres, and in their angular orientation. Not surprisingly the smectics show pronounced optical activity.

Various intermediate mesogenic stages may develop. For example, 4,4'-di-*n*-heptyl-oxyazoxybenzene shows the following intermediate stages between solid and isotropic liquid:

$$\text{solid} \xrightarrow{\ \ \ \ } \underset{74\,°C}{\text{smectic C}} \xrightarrow{\ \ \ \ } \underset{95\,°C}{\text{nematic}} \xrightarrow{\ \ \ \ } \underset{124\,°C}{\text{isotropic}}$$

This order, incidentally, also demonstrates the relative degrees of disorder characterizing the various intermediate mesophases.

The Nematic Mesophase

An idealized model of a nematogen is shown in Figure 9.1: it is schematic to the extent that perfect molecular alignment is implied, when in fact thermal fluctuations of r.m.s. amplitude up to 40° may occur. The random distribution of molecular centres, subject to simple geometric constraints, is essentially liquid-like, whilst the local degree of orientational order is more appropriate to a

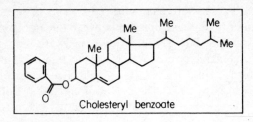

Cholesteryl benzoate

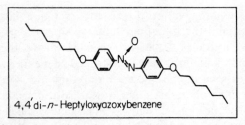

4,4′di-*n*-Heptyloxyazoxybenzene

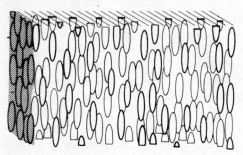

Figure 9.1 Molecular organization in the nematic mesophase

solid, and we describe this local degree of preferred orientation by the *director* **n(r)**, a unit vector field which is parallel to the local preferred orientation and varies from point to point.

The specification of the local degree of orientational order is given in terms of the *Zwetkoff parameter*, S, defined by the mean of the angle θ between the symmetry axis and the director:

$$S = \langle \tfrac{1}{2}(3\cos^2\theta - 1)\rangle \tag{9.1}$$

where $\langle \ \rangle$ represents a time or ensemble average over the molecular distribution.

S turns out to be the second unassociated Legendre function $P_2^0(\cos\theta)$: this definition is convenient in that it adopts the values $S = 1$ for perfect orientational alignment, and $S = 0$ for perfect isotropy. Values of S as high as 0·8 have been recorded, but typically it varies from $\sim 0·4$, falling with increasing temperature and thermal disordering, discontinuously dropping to zero at the nematic–isotropic transition temperature, T_{trans}. At this temperature the optical activity and turbidity vanishes—hence the earlier name of 'clearing point' for the temperature T_{trans}.

At present this orientational melting is discussed in terms of molecular field theory in which the potential energy of a representative molecule is calculated in the mean field of its neighbours. Such a model does not account for the reaction of the molecule on the environment. Nevertheless, many of the principal features are exposed in this approach. Maier and Saupe (1958) proposed the anisotropic pairwise intermolecular potential

$$\Phi_{12} = u(r)(3\cos^2\theta_{12} - 1)/2 \tag{9.2}$$

where $u(r)$ is the purely radial component of the interaction, generally attributed to the London dispersion forces, and where θ_{12} is the relative orientation of molecules 1 and 2. The radial component will, of course, show an r^{-6} dependence working only to first order, and in fact

$$u(r) = -\frac{3hv\Delta\alpha^2}{20r^6} \tag{9.3}$$

where $\Delta\alpha$ is the anisotropy in the molecular polarizability $(\alpha_{\parallel} - \alpha_{\perp})$. If now we average over all configurations (spatial and orientational) of molecule 2, we obtain the potential of molecule 1 in the mean field approximation as

$$\Phi_1(\cos\theta_1) = \bar{u}S(\tfrac{3}{2}\cos^2\theta_1 - \tfrac{1}{2}) \tag{9.4}$$

where S is yet to be determined, and θ_1 is the angle between the long molecular axis and the director, which we shall take to lie along the z-axis. The molecular orientational distribution is then given by the single-particle Boltzmann expression

$$f_1(\cos\theta) = \exp\left\{-\frac{\bar{u}S}{kT}(\tfrac{3}{2}\cos^2\theta - \tfrac{1}{2})\right\} \tag{9.5}$$

The single-particle potential Φ_1, then follows directly as

$$\Phi_1(\cos\theta) = \frac{\iint \Phi_{12}(\cos\theta_{12})f_1(\cos\theta_2)\,\mathrm{d}2\,\widehat{\mathrm{d}2}}{\iint f_1(\cos\theta_2)\,\mathrm{d}2\,\widehat{\mathrm{d}2}} \tag{9.6}$$

where $\iint \mathrm{d}2\,\widehat{\mathrm{d}2}$ represents integration over all the spatial and orientational configurations of molecule 2. It follows immediately from (9.6) that

$$\Phi_1(\cos\theta_1) = \bar{u}(\tfrac{3}{2}\cos^2\theta_1 - \tfrac{1}{2})\langle(\tfrac{3}{2}\cos^2\theta_2 - \tfrac{1}{2})\rangle_f \tag{9.7}$$

where $\langle\ \rangle_f$ represents the thermodynamic average of the argument over the distribution $f_1(2)$ (equation (9.5)). Self-consistency between equations (9.4) and (9.7) requires that

$$S = \langle (\tfrac{3}{2} \cos^2 \theta - \tfrac{1}{2}) \rangle_f \qquad (9.8)$$

which, in fact, we understood from (9.1).

This is the fundamental equation of the Maier–Saupe mean-field theory of the nematic phase, and can be solved to find the variation of the order parameter S with temperature (Figure 9.2).

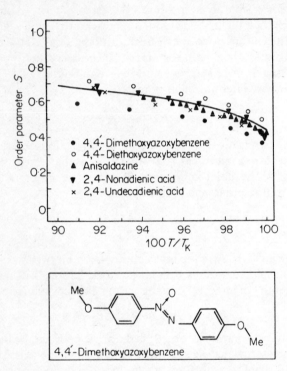

Figure 9.2 Variation of the nematic order parameter S with reduced temperature in the mean-field approximation. Points show experimental data for various nematic systems

Φ_1 is a *pseudopotential* attributable to orientational effects only, and clearly vanishes in an isotropic ($S = 0$) fluid. The excess Helmholtz free energy attributable to orientational effects may be calculated on the basis of (9.4), and is found to vanish when

$$\frac{\bar{u}}{kT} = -4.542 \qquad (9.9)$$

and this defines the nematic–isotropic transition temperature. Equation (9.4) is obviously expressible in terms of the transition temperature, and so we might expect a number of nematic substances plotted in reduced temperature co-ordinates (T/T_{trans}) to exhibit the same qualitative behaviour. A comparison is made in Figure 9.2 where the agreement over a variety of substances is seen to be quite good, although we haven't attempted to incorporate the small discontinuous variation of $\bar{u}$ with the density change (~ 5 per cent) across the nematic–isotropic transition.

Some preliminary attempts have been made recently to apply the BGY and PY integral techniques to liquid-crystal systems. Approximation is made in terms of a superposition expression for $g_{(3)}(\mathbf{1, 2, 3}, \widehat{\mathbf{1}}, \widehat{\mathbf{2}}, \widehat{\mathbf{3}})$ or an angular direct correlation $c(\mathbf{12}, \widehat{\mathbf{1}}, \widehat{\mathbf{2}})$ but we cannot discuss these here.

The director will vary from point to point in the nematic phase, having a typical spatial coherence of several hundred angstroms. The diamagnetic susceptibility of the molecules will, however, cause the director to align in an external magnetic field, as we indicated in the introduction, though we should emphasize that the magnetic field aligns the *director*, and makes no difference to the molecular order parameter S. The director aligns and saturates in a magnetic field of several kilogauss, although a nematic aligned between two parallel plates will not necessarily undergo an orientational transition in a magnetic field. This *Fréedericksz transition* will only occur if the separation of the plates satisfies the condition

$$\frac{d}{\pi} \geqslant \xi$$

where ξ is the *coherence length*. With alignment of the director, the nematogen behaves as a uniaxial crystal, and the dielectric anisotropy $\Delta\varepsilon$ is then at its most pronounced. In Figure 9.3 we show the development of $\Delta\varepsilon = (\varepsilon_\parallel - \varepsilon_\perp)$ with increasing magnetic field for 4,4-dimethoxyazobenzene.

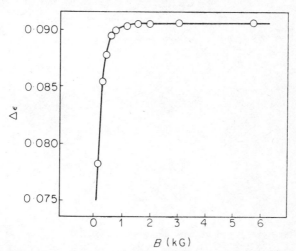

Figure 9.3 Dielectric anisotropy for the mesophase of 4,4′-dimethoxyazoxybenzene at 134·7 °C as a function of magnetic field strength

There are also mechanical methods of aligning the director. A nematogen can, for example, be aligned parallel to a glass plate, and if the plate has been previously rubbed in one direction, the alignment is parallel to both the plate and to the direction in which the plate was rubbed. Again, at the liquid surface there is likely to be pronounced alignment of the director, and this must certainly be incorporated in any theory of the surface tension. It should be pointed out, however, that the origins of these surface effects are still largely unknown.

The director will vary continuously from point to point, and the characterizing vector field will show features generally designated as *splay*, *twist* and *bend*. With all these distortions there will be an associated free-energy density $F(\mathbf{r})$ which, like the director field itself, will be a function of position $\mathbf{r}$. The three distortions are shown qualitatively in Figure 9.4 and the free energy associated with the distortions of the director field $\mathbf{n}(\mathbf{r})$ may be expressed in terms of

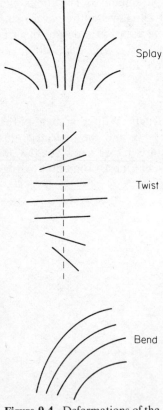

Figure 9.4 Deformations of the director field: splay, twist and bend associated with the elastic constants K_{11}, K_{22} and K_{33} respectively

divergence and curl operations:

$$F(\mathbf{r}) = \tfrac{1}{2}\{\underset{\text{splay}}{K_{11}(\nabla \cdot \mathbf{n})^2} + \underset{\text{twist}}{K_{22}[\mathbf{n} \cdot (\nabla \times \mathbf{n})]^2} + \underset{\text{bend}}{K_{33}[\mathbf{n} \times (\nabla \times \mathbf{n})]^2}\} \quad (9.10)$$

where the coefficients K_{11}, K_{22} and K_{33} are the elastic constants. (This kind of expression of the free-energy density in a distorted body is familiar in the discussion of elastic properties of anisotropic solids). The elastic constants are very small ($\sim 10^{-6}$ dyne) in comparison to the orientational free energy in the liquid crystal, and so it is an easy matter to distort the director. Indeed, the smallest perturbations due to impurities or surface inhomogeneities are sufficient to establish long-range 'dislocation' or singularities in the director field, termed *disinclinations*. Point and line disinclinations of various kinds may develop, and have associated high distortion energy densities.

The full technological impact of the liquid-crystal systems has not yet been reached, but already the application of nematogens in low-consumption display devices is sufficient to warrant the current intense research activity into these systems. A surface-aligned nematogen between two transparent electrodes is quite transparent, as shown in Figure 9.5. The application of a small potential

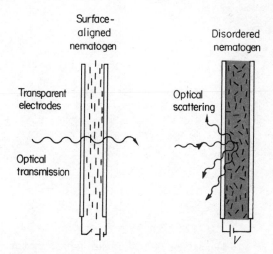

Figure 9.5 Transmission of optical wavelengths through a nematic sandwich. Application of a small potential disorders the nematogen sufficiently to become opaque at optical wavelengths

across the electrodes, no more than $\sim 5V$ in some cases, is sufficient to cause optical scattering, and although the mechanism is not yet fully understood, it is thought to be associated with the disordering effect of ionic impurities drifting under the effect of the electric field. If, of course, the electrodes were comprised of segments in a digital pattern, application of a potential to the appropriate segments will provide a low-consumption display device, and indeed, these are

212

commercially available. One of the problems however is the response and decay times of the perturbed mesophase.

Another application of this kind of device is in the self-tinting glass where the applied field is governed by the intensity of the incident radiation. Clearly, these opto-electric devices are of immense importance in the conversion of mechanical perturbations into opical signals.

Cholesteric Liquid Crystals

Many esters of cholesterol, though not cholesterol itself, exhibit a 'twisted nematic' mesophase in which the director describes a helix. Thus, the long axes of the molecule, though parallel, are rotated by $\sim 15'$ from layer to layer (Figure 9.6). The twisting is attributed to the molecular structure which does not permit

Figure 9.6 Twisted nematic form of the cholesteric mesophase

superposition, but only a slow structural rotation. As we might expect, such a mesogen is likely to show pronounced optical activity, and indeed, one milli-meter thickness of a uniformly oriented cholesteric is sufficient to rotate the plane of polarization through 60,000°. This is to be compared with organic compounds depending upon molecular assymmetry for their optical activity which can manage only $\sim 300°$.

The unperturbed pitch Z_0 of the helix can be determined experimentally from the Bragg condition

$$\lambda_0 = 2Z_0 \sin \theta \tag{9.11}$$

and Z_0 may be $\sim 3000\text{Å}$ in which case the interference effects may be observed at optical wavelengths, and indeed account for the beautiful colours often displayed by thin films of these liquid systems in selective reflection.

The application of a magnetic field perpendicular to the axis of the helix has the effect of aligning the director, whilst the fundamental molecular structure tries to maintain the natural helicity Z_0. The result is that the pitch of the helix Z increases with increasing field until it diverges logarithmically, i.e. becomes nematic, at very large field strengths, (~ 10 k gauss) the periodic structure having disappeared completely (Figure 9.7).

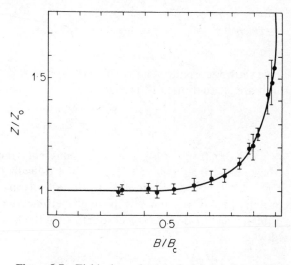

Figure 9.7 Field dependence of the relative pitch Z/Z_0 of the cholesteric phase formed on adding cholesteryl acetate to 4,4′-dimethoxyazoxybenzene at 119 °C

The more general expression for the energy density is now (cf. equation (9.10))

$$F(\mathbf{r}) = \tfrac{1}{2}\left\{ K_{11}(\nabla \cdot \mathbf{n})^2 + K_{22}\left[\mathbf{n} \cdot (\nabla \times \mathbf{n}) + \frac{2\pi}{Z_0} \right]^2 + K_{33}[\mathbf{n} \times (\nabla \times \mathbf{n})]^2 \right\} \quad (9.12)$$

where the extra term $(2\pi/Z_0)$ in the twist ensures that the free energy is a minimum when the director forms a helix of natural pitch Z_0.

We may calculate the critical field B_c required to establish a nematic mesophase in a cholesteric using the free-energy density concepts outlined above.

Imagine the cholesteric director configuration shown in Figure 9.8. The director at 0 is located in the coordinate frame by the director components

$$n_x = 0, \qquad n_y = 0, \qquad n_z = 1 \qquad (0, 0, 1) \qquad (9.13)$$

whilst at P the director has rotated to $(0, 1, 0)$. At an intermediate point we have

$$n_x = 0, \qquad n_y = \cos \phi, \qquad n_z = \sin \phi \qquad (9.14)$$

214

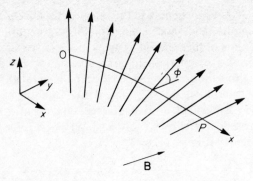

Figure 9.8 Alignment of director by magnetic forces

where ϕ is the angle shown in Figure 9.8. Provided there is no splay or bend, the free energy per unit area (equation (9.12)) reduces to

$$F(x) = \frac{1}{2}\left\{ K_{22}\left[\left(\frac{\mathrm{d}\phi}{\mathrm{d}x}\right) - \frac{2\pi}{Z_0}\right]^2 - F_{\mathrm{mag}} \right\} \qquad (9.15)$$

since $\mathbf{n}\cdot(\nabla \times \mathbf{n}) \to (\mathrm{d}\phi/\mathrm{d}x)$ from (9.14), and F_{mag} represents the free-energy density due to the applied magnetic field B. It is not difficult to show that $F_{\mathrm{mag}} = \Delta X(\mathbf{B}\cdot\mathbf{n})^2 = \Delta X B^2 \cos^2\phi$ where ΔX is the anisotropy in the diamagnetic susceptibility $(X_{\parallel} - X_{\perp})$ and ϕ is the angle between the field and the symmetry axis. Integrating (9.15) with respect to x to yield the total free energy, this is minimized when

$$K_{22}\left(\frac{\mathrm{d}^2\phi}{\mathrm{d}x^2}\right) + \Delta X B^2 \sin\phi \cos\phi = 0 \qquad (9.16)$$

The pitch of the helix is then

$$Z = \int_0^{2\pi} \frac{\mathrm{d}x}{\mathrm{d}\phi}\,\mathrm{d}\phi \qquad (9.17)$$

which is obtained as a single integral of (9.12), and the *relative* pitch follows as

$$\frac{Z}{Z_0} = \left(\frac{2}{\pi}\right)^2 K(k)E(k) \qquad (9.18)$$

where

$$\frac{k}{E(k)} = \frac{Z_0 B}{\pi^2}\left(\frac{\Delta X}{K_{22}}\right)^{\frac{1}{2}} \qquad (9.19)$$

and $E(k)$ and $K(k)$ are elliptic integrals. For a nematic phase the critical field B_{c} is finally given as

$$B_{\mathrm{c}} = \frac{\pi^2}{Z_0}\left(\frac{K_{22}}{\Delta X}\right)^{\frac{1}{2}} \qquad (9.20)$$

We finally observe that the cholesteric–nematic transition is not a true phase transition since there is no discontinuity in the free energy at the critical field, B_c.

The helical pitch of cholesteric systems, and hence their selective optical properties, are also sensitive to thermal, mechanical and electrical perturbation and have been used to detect hot spots and hence flaws in integrated circuits, and even in the location of tumours in the human body. By appropriate cholesteric multicomponent mixtures it is possible to pass through the entire optical range of selective reflection over temperature variations as little as 1 °C, or as much as 50 °C.

Smectic Liquid Crystals

The smectogens are most similar to conventional crystalline systems showing a highly developed spatial and orientational molecular organization (Figure 9.9).

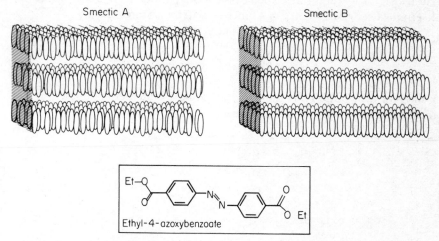

Figure 9.9. Molecular organization in the smectic A and smectic B mesophases

The distinguishing feature of the smectics is the essentially *lamellar* structure which is stabilized by strong interactions between certain *regions* of the molecule: the various morphologies within the lamella classifying the smectogen into one of the categories A through H. These liquid crystals are distinguished from the solid by the ability of the layers to slide or rotate over one another.

Only the smectics A, B, C and D have been identified with any certainty, and phases A and B are shown in Figure 9.9. Phase A corresponds to a layered nematic configuration whilst in C the direction of the molecules is tilted with respect to the layers. In smectophase B the molecular arrangement is believed to be highly ordered, perhaps hexagonal-close-packed, whilst D is known to be cubic. McMillan (1971) has extended the Maier–Saupe nematic mean field analysis to smectic A systems by assuming that in addition to being preferentially oriented along the z-axis, there is layering in the z-direction (Figure 9.9) such that the single-particle potential felt in the mean field of its neighbours is

$$\Phi_1(z_1 \cos \theta) = \bar{u}S\left[1 + \alpha \cos\left(\frac{2\pi z}{d}\right)\right](\tfrac{3}{2}\cos^2\theta - \tfrac{1}{2}) \qquad (9.21)$$

where both translational and orientational effects are now included. α is a measure of the interaction strength ($\alpha = 0$ corresponding to the nematic case) and d is the length of the molecule. Now, although the minimum energy configuration corresponds to the molecular centre-of-mass lying on one of the planes with its long axis in the z-direction, there will be thermal fluctuations in the orientation and location in the centre-of-mass so that a second order parameter has now to be introduced into the single-particle potential:

$$\Phi_1(z, \cos \theta) = \bar{u}S\left[1 + \sigma\alpha \cos\left(\frac{2\pi z}{d}\right)\right](\tfrac{3}{2}\cos^2\theta - \tfrac{1}{2}) \qquad (9.22)$$

The density wave is now described by the order parameter, and σ and S are to be determined self-consistently as a function of temperature

The single-particle distribution function is, in Boltzmann form

$$f_1(z, \cos \theta) = \exp\left(-\Phi_1(z, \cos \theta)/kT\right) \qquad (9.23)$$

Using this distribution function and the two-body interaction function we may recalculate the single-particle potential:

$$\Phi_1(z_1, \cos \theta_1) \equiv \frac{\iint \Phi_{12}(r_{12}, \cos \theta_{12})f_1(z_2, \cos \theta_2)\,\mathrm{d}2\,\mathrm{d}\hat{2}}{\iint f_1(z_2, \cos \theta_2)\,\mathrm{d}2\,\mathrm{d}\hat{2}}$$

$$= \bar{u}\left[(\tfrac{3}{2}\cos^2\theta_1 - \tfrac{1}{2})\langle\tfrac{3}{2}\cos^2\theta_2 - \tfrac{1}{2}\rangle_f\right.$$

$$\left. + \alpha \cos\left(\frac{2\pi z_1}{d}\right)(\tfrac{3}{2}\cos^2\theta_1 - \tfrac{1}{2})\left\langle \cos\left(\frac{2\pi z_2}{d}\right)\langle\tfrac{3}{2}\cos^2\theta_2 - \tfrac{1}{2}\rangle_f\right\rangle\right]$$

$$(9.24)$$

Self-consistency between (9.22) and (9.24) requires that

$$S = \langle(\tfrac{3}{2}\cos^2\theta - \tfrac{1}{2})\rangle_f \qquad (9.25)$$

$$\sigma = \left\langle \cos\left(\frac{2\pi z}{d}\right)(\tfrac{3}{2}\cos^2\theta - \tfrac{1}{2})\right\rangle_f \qquad (9.26)$$

where the z average implied in σ is taken over one density wavelength, i.e.

$$\int_0^d \mathrm{d}z$$

The Smectic-A–Nematic–Isotropic Phase Transition

McMillan has solved the coupled equations (9.25) and (9.26) self-consistently for the order parameters S and σ. S, introduced by Maier and Saupe in the mean-field approximation for nematic systems, describes the orientational order whilst σ describes the amplitude of the density wave.

There are three types of solution:

(i) $\sigma = S = 0$: no order, characteristic of the isotropic phase.

(ii) $\sigma = 0, S \neq 0$: orientational order only, the theory reduces to the Maier–Saupe theory of the nematic phase.

(iii) $\sigma \neq 0, S \neq 0$: orientational and translational order, characteristic of the smectic A phase.

To determine which of the three phases—smectic A, nematic or isotropic—is stable at a given temperature we have to calculate the free energy of the system:

$$F = -kT \ln Z_N$$

where Z_N is the N-body partition function:

$$Z_N = N \ln \left\{ d^{-1} \int_0^d \int_0^1 \exp \left\{ -\frac{\bar{u}}{kT} \left[S + \sigma\alpha \cos\left(\frac{2\pi z}{d}\right) \right] (\tfrac{3}{2}\cos^2\theta - \tfrac{1}{2}) \right\} d(\cos\theta)\,dz \right\}$$

$$= N \ln Z_1$$

where Z_1 is the single-particle partition. The internal energy U is simply the thermodynamic average over the two-particle interaction:

$$U = -\tfrac{1}{2}\bar{u}(S^2 + \alpha\sigma^2)$$

and the entropy is given from $F = U - TS$. Finally, the specific heat at constant volume is

$$C_v = T\left(\frac{\partial S}{\partial T}\right)_v$$

The phase diagrams may now be shown for various values of the parameter corresponding to the amplitude of the density wave. The value of α may, in fact, be related to the chain length, and as we shall see the smectic-A–nematic transition temperature rises with α and increasing chain length, and eventually meets the nematic–isotropic temperature. For large chain lengths we have only the smectic A and isotropic phases.

In Figure 9.10 we show the three phases ($\alpha = 0.6$) as a function of the reduced temperature (T/T_{NI}) where T_{NI} is the nematic–isotropic transition temperature. We see from the continuity of the entropy at the smectic–nematic transition that this is of second order, whilst the nematic–isotropic transition is first order. In Figure 9.11 the smectic–nematic transition is now first-order ($\alpha = 0.85$) and both S and σ show discontinuities at T_{SN}.

218

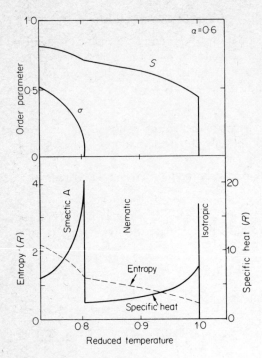

Figure 9.10 Order parameters S and σ, entropy and specific heat, versus reduced temperature for the theoretical model with $\alpha = 0.6$ showing the second-order smectic-A–nematic transition and the first-order nematic–isotropic transition

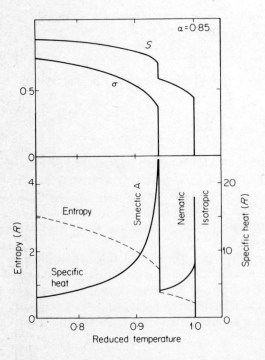

Figure 9.11 Order parameters, entropy and specific heat against reduced temperature for $\alpha = 0.85$ showing the first-order smectic-A–nematic transition

Evidently the more pronounced density wave sustains the smectic A phase to a higher temperature. For a sufficiently large value of the density-wave parameter ($\alpha = 1\cdot1$, Figure 9.12) we see that the smectophase may be sustained until orientational rather than translational order is thermally disrupted. Of

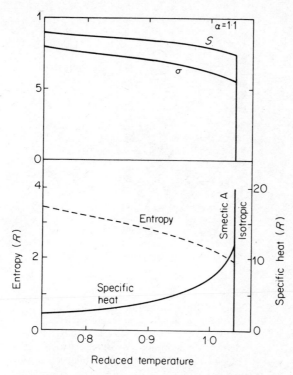

Figure 9.12 Order parameters, entropy and specific heat against reduced temperature for $\alpha = 1\cdot1$ showing the first-order smectic-A–isotropic transition

course, with the breakdown of orientational order, the loss of translational order is implicit: this coupling of the orientational and translational order is evident both physically and from equation (9.26). Under these circumstances nematophase develops and we have a first-order transition between the smectic A and isotropic phases.

These results are certainly in qualitative agreement with experiment, and these molecular processes undoubtedly underlie the thermodynamic features of the smectic-A–nematic–isotropic transition. From the thermodynamic point of view, the nematic phase may be extended to include the cholesterics since the small free energy associated with the cholesteric twist is negligible in comparison with the other thermodynamic quantities.

220

Molecular Model for Phase Transitions in Biological Membranes

Diffusive transport through the cell membrane and other membrane functions are believed to be regulated by the molecular configuration at the cell boundary (Figure 9.13), in particular by means of a phase transition of the lipid molecules.

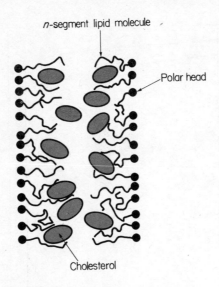

Figure 9.13 Cell membrane composed of lipid molecules with embedded cholesterol molecules. Transport across the membrane may be regulated by phase transitions of the lipid bilayer

Unlike the classical rigid liquid-crystal molecules, the lipids may be regarded as being statistically equivalent to a small chain of elements with no restrictions on their possible orientations. The number of independent segments for a given molecule is determined by comparison with experiment.

We assume the usual Maier–Saupe anisotropic interaction between molecular *segments*, and in the mean field approximation we obtain the usual single-particle potential:

$$\Phi_1 = \bar{u}(\tfrac{3}{2}\cos^2\theta - \tfrac{1}{2})S \tag{9.27}$$

where θ is the angle between the long axis of the segment and the preferred axis of orientation.

There are *additional* orienting agents operating within the cell membrane besides the anisotropic interaction indicated above. For example, cholesterol molecules within the bilayer restrict the orientational freedom and impose a degree of order on the chains. Near the planes defined by the polar heads the chains cannot have a random configuration for purely geometrical reasons. Marčelja has proposed incorporating these additional ordering effects in terms of a single *constant* parameter, η_0, to be added to the molecular field. The single-particle potential then becomes

$$\Phi_1(S, \eta_0) = \bar{u}(\tfrac{3}{2}\cos^2\theta - \tfrac{1}{2})(S + \eta_0) \tag{9.28}$$

where η_0 has the role of *enhancing* the order in the lipid membrane. The order parameter S is now determined so as to be consistent with the anisotropic pair interaction. Thus, as before, the molecular distribution is

$$f_1 = \exp\left(-\Phi_1(S, \eta_0)/kT\right) \tag{9.29}$$

in Boltzmann form. The order parameter follows as

$$S = \langle(\tfrac{3}{2}\cos^2\theta - \tfrac{1}{2})\rangle_{f_1} \tag{9.30}$$

The per particle entropy is given by the free-energy relationship

$$TS = n\bar{u}S^2 + nkT\ln\int_0^1 f_1(S, \eta_0)\,d(\cos\theta) \tag{9.31}$$

where n is the number of independent segments for the molecule.

The variation of the order parameter S with temperature is shown for two constant values of η_0 in Figure 9.14. For $\eta_0 = 0.0174$ the order parameter

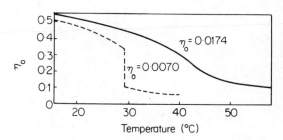

Figure 9.14 Order parameters for $\eta_0 = 0.0174, 0.0070$

varies continuously with temperature, whilst for $\eta_0 = 0.0070$, the membrane undergoes a first-order transition, Figure 9.14: the critical value separating these two regions is $\eta_0' = 0.0103$. Both these processes have been observed experimentally. Natural membranes tend to show more of a continuous transition, presumably attributable to the hindering effect of the cholesterol molecules within the bilayer, whilst in artificial membranes where the molecular segments have a greater degree of freedom, transition is more clearly first order.

A comparison of the specific heat of pronase-treated *Mycoplasma laidlawii* membrane and the theoretical value given by $C_v = T(\partial S/\partial T)_v$ is shown in Figure 9.15 for $\eta_0 = 0.0174$. From a comparison of the experimental and theoretical transition entropies ($\sim 3.6 \times 10^3$ cal mol^{-1}) we conclude that the

222

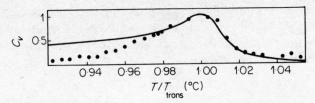

Figure 9.15 Specific heat of pronase-treated *Mycoplasma laidlawii* membranes as a function of temperature taking $\eta_0 = 0.0174$. Solid points represent experimental data

number of chain segments $n \sim 7.7$, which is a reasonable value for a typical lipid molecule with thirty or forty CH_2 units.

References

G. Luckhurst, Review article in *Physics Bulletin* (May 1972).
P. de Gennes, *The Physics of Liquid Crystals*, Oxford (1974).
S. Marčelja, *Nature*, **241**, 451 (1973); *Biochimica et Biophysica Acta*, **367**, 165 (1974).

CHAPTER 10

Numerical Methods in the Theory of Liquids

Introduction

We have already encountered some results of the machine calculations in earlier chapters in which they provided 'experimental' data for idealized systems of particles—systems interacting through hard-sphere and square-well potentials. For reasons of mathematical expediency we were particularly concerned with these caricatures of realistic interactions in testing the various approximate integral equations, not only because of their fundamental simplicity but also because comparison of theory with experiment for realistic interactions is obscured by uncertainty in the pair potential operating in real systems. In the case of the liquid inert gases, for example, is the theoretical discrepancy to be attributed to the experimental scattering data, its inversion to yield the distribution function, the assumed form of the pair potential, or a combination of all these uncertainties? Here the machine calculations again play an important role in simulating a realistic system interacting through a *prescribed* pair potential, and although it may differ in certain respects from the real interaction, it does enable direct comparison with theory to be made and eliminates several of the intervening uncertainties.

Alternatively, we may attempt to simulate the thermodynamic properties of realistic systems as closely as possible in an inverted sense, and investigate the role of many-body effects on the assumed form of pair potential.

The wealth of microscopic data generated by such an 'experiment' provides an unambiguous basis for the testing and refinement of both the structural and kinetic theories of fluids. Indeed, as we have pointed out, the machine simulations have now reached a level of sophistication such that we may temporarily abandon comparison with experiment and instead study idealized systems which are simple enough to aid theoretical developments.

Essential Limitations of the Computer Schemes

We have to remember the inherent limitations of computer simulations, in particular the small number of degrees of freedom which may be dealt with on even the largest conceivable computer. In the case of spherically symmetrical, structureless particles the current restriction on the number of particles is $N \sim 10^3$. For more complicated molecules such as water in which rotational and vibrational degrees of freedom have to be incorporated, assemblies of no more than $\sim 10^2$ particles are currently feasible. The extent to which the thermodynamic properties of such microscopic assemblies can be considered representative of an infinite system constitutes a limitation of the technique—a system of

224

10^3 particles represents a cubical array of side ~ 10 atomic diameters at liquid densities, and a statistical error is engendered by the fact that only a finite number of particles are used. An estimate of the error can be obtained by comparing calculations based on the same parameters using different numbers of particles.

In order to simulate as closely as possible the behaviour of an infinite system 'periodic boundary conditions' are generally used (Figure 10.1). A particle A

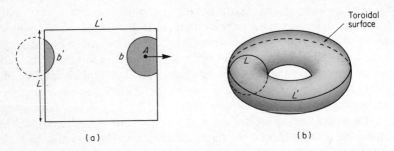

Figure 10.1 (a) Schematic two-dimensional array showing periodic boundary conditions. The range of interaction of particle A is indicated by the shaded region. Clearly, the dimension of the array must be such that no bb'-interaction can result. (b) Two-dimensional toroidal surface formed from the rectangular array in (a) by implementing the periodic boundary conditions

leaving one face of the cell reappears through the opposite face: obviously the dimension of the fundamental cell and the range of the interaction (shown by the shaded area in Figure 10.1a) must be such that there is no bb' correlation. Both for this and reasons of computational expediency the interaction is truncated beyond a certain radius, particle A taking account of all particles within the truncation radius shown by the shaded area. For a three-dimensional system there will be twenty-six 'images' of the fundamental cell and to this extent boundary effects are minimized but not eliminated, and the statistical error engendered in a finite assembly is in no way reduced by this device.

Since the maximum temporal extent for which a dynamical event can be followed without inordinate expenditure of computer time is $\sim 10^{-11}$ s, we are clearly limited to microscopic non-hydrodynamic events having a relaxation time substantially less than 10^{-11} s.

Phase transitions and critical phenomena are particularly difficult to follow in these simulation schemes: the long-wavelength cooperative phenomena associated with these effects are inhibited by the cell dimension and consequently the critical temperature, for example, is overestimated by ~ 7 per cent, and the specific heat is underestimated—both effects attributable to the suppression of fluctuations in the assembly.

The Molecular Dynamics Method

In this method the classical Newtonian equations of motion of an assembly of particles are solved numerically and integrated to yield the evolution of the configurational and velocity distributions. The particles are released from an arbitrary (non-overlapping) configuration within the fundamental cell either from rest or with a random distribtion of velocities. The system evolves rapidly, attaining a Maxwellian velocity distribution after about six collisions per particle in the case of hard spheres in which the energy of the assembly is purely kinetic—rather more for realistic interactions in which configurational adjustment must accompany the potential–kinetic energy exchange in attaining the equilibrium velocity distribution.

The position and velocity coordinates are stored as a function of time, and after the transient effects of the initial starting configuration have died out the data may be analysed for the microstructure and microkinetics of the assembly. The net force on an atom i may be determined classically by taking the vector sum over the neighbouring particles (Figure 10.2)

$$- \sum_{j}^{N-1} \nabla\Phi(ij) \tag{10.1}$$

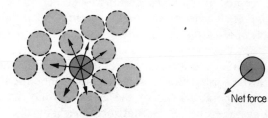

Figure 10.2 Calculation of net force on a molecule by its environment

presupposing a knowledge of the distribution of atomic centres and the interaction potential $\Phi(ij)$. If the configurational and momentum coordinates of each atom at some time t are stored, then the classical trajectory of atom i over a period of time Δt may be determined by solution of the Newtonian equation as a sequence of discrete linear steps. The new position coordinate in the absence of any external fields will be

$$\mathbf{q}_i(t + \Delta t) = \mathbf{v}_i(t)\Delta t - \frac{1}{2m_i} \sum_{j}^{N-1} \nabla\Phi(ij)(\Delta t)^2 + \mathbf{q}_i(t) \tag{10.2a}$$

whilst the new velocity coordinate will be

$$\mathbf{v}_i(t + \Delta t) = \mathbf{v}_i(t) - \frac{1}{m_i} \sum_{j}^{N-1} \nabla\Phi(ij)(\Delta t) \tag{10.2b}$$

The trajectory of atom i therefore traces out a path in space similar to that shown in Figure 3.7. Short-lifetime vibratory modes A are interspersed with diffusive motions, and the phonon spectrum of a liquid shows these two components quite clearly. The simulation proceeds by sequentially determining the net force and trajectory of each of the N particles in the assembly and the system executes an implicitly pairwise evolution.

The conflicting interests of computational expediency and significance of the calculated result restrict the time graining $\Delta t \sim 10^{-14}$ s. If coarser time graining is used, overlapping configurations of atoms will develop and the particles will subsequently separate with infinite or near-infinite velocities depending upon the details of the repulsive core of the interaction potential.

The Monte Carlo Method

The term Monte Carlo has come into use to designate numerical methods in which specifically stochastic elements are introduced in contrast to the completely deterministic algebraic expressions of the molecular dynamics approach. The particular form used in liquid state physics is that devised by Metropolis *et al.*

The Monte Carlo method consists in generating a set of molecular configurations by random displacements of the N particles in the model. A configuration N_j is accepted or rejected according to a criterion *which ensures that in the limit of an infinite number of transitions a given configuration occurs with a probability proportional to the Boltzmann factor* $\exp(-\Phi_N/kT)$ *for that configuration.*

We need a criterion for the acceptability of the new configurations such that they develop with the correct Boltzmann weighting: then the thermodynamic and configurational characteristics of the assembly may be determined as actual canonical averages in the true statistical mechanical sense:

$$\langle f \rangle = \int f(x)P(x)\,\mathrm{d}x \bigg/ \int P(x)\,\mathrm{d}x \tag{10.3}$$

The thermodynamic average $\langle f \rangle$ is thus determined as a weighted phase average where $P(x)$ is the weighting function. A commonly used prescription for the acceptability of a trial configuration is the following: a particle of the system is selected, either serially or at random, and given a random displacement from state i to state j. Suppose the increase in the total configurational energy is $\Delta\Phi_N^{ij}$, then if $\Delta\Phi_N^{ij}$ is negative the move is accepted and the new configuration replaces the old one—the new configuration is evidently more probable than the former one. If $\Delta\Phi_N^{ij}$ is positive, however, the move is accepted only with the probability $P_{ij} = \exp(-\Delta\Phi_N^{ij}/kT)$: the machine then selects a random decimal number in the range 0 to 1 and compares it to $\exp(-\Delta\Phi_N^{ij}/kT)$. If the exponential is the greater the move is allowed, otherwise the move is rejected. If a move is rejected the previous configuration is counted again. The distribution in configuration space develops as the Boltzmann factor $\exp(-\Phi_N/kT)$. The overall chain average of any function therefore converges to the canonical ensemble average as the number of steps in the chain $\to \infty$.

Comparison of the Molecular Dynamics and Monte Carlo Methods

Many of the essential limitations of the computer schemes discussed earlier apply to both the principal simulation methods. The restriction to small, finite assemblies is inevitable in both procedures, and periodic boundary conditions are introduced to eliminate surface effects. In neither case is pairwise additivity of the potential function an essential restriction, although it is generally employed because of the great simplification that ensues.

The great advantage of the molecular dynamics approach lies in its ability to deal with non-equilibrium and transport phenomena, provided the relaxation time for the process is significantly smaller than the computation time. Since the maximum temporal extent for which a dynamic event can be followed is $\sim 10^{-11}$ s, we are clearly limited to microscopic non-hydrodynamic events.

The equivalence of the asymptotic results of the Monte Carlo configurational averages for the thermodynamic quantities and the molecular dynamics results obtained as time averages over phase-space trajectories depends essentially on the passage to the asymptotic limit $N \to \infty$, $V \to \infty$, $N/V \to$ constant. The two methods should produce results in agreement to order N^{-1}, and one would generally expect agreement of the two approaches within statistical error.

One particular advantage of the Monte Carlo method is that it may be relatively easily extended to quantum-mechanical systems in which the exchange symmetry of the single-particle wavefunctions must be preserved.

Hard Discs and Spheres

The molecular dynamics simulation of an assembly of rigid discs is shown in Figure (2.5b) together with the Ree–Hoover Padé (3–3) approximant to the equation of state using the hard-disc virial coefficients B_2 to B_6. (equation (2.48)). In the case of hard discs and hard spheres the development of overlapping configurations is inevitable, but provided the time step is small enough the many-body interaction can be resolved as a series of two-body encounters. When a contact or overlap configuration occurs the centres separate classically with a conservation of momentum, and infinite velocities do not develop.

The Alder and Wainwright result shows the familiar van der Waals loop linking the fluid and solid branches of the isotherm, and in Figure 6.9 we show configurations taken from Wood's hard-disc Monte Carlo isotherm. Figure 6.9(a) is associated with the 'fluid' locus whilst Figure 6.9(b) is taken from the 'solid' locus of the equation of state: the 'crystallography' is seen to be markedly different in the two cases.

The development of the van der Waals loop instead of a horizontal tie-line linking the two branches of the isotherm warrants some comment (see Chapter 5): although the tie-line may be located by the Maxwell equal-areas rule, Alder

228

and Wainwright suggest that for finite systems the cell dimension inhibits the structural rearrangement which would be consistent with minimum free energy, and consequently metastable extensions of the predominant phase occur. We have seen in Chapter 5 that the melting transition is characterized by long-wavelength cooperative phenomena, and we can understand that the finite molecular dynamic and Monte Carlo assemblies will have difficulty in following the transition.

The computational details involved in the study of the hard-disc system are very simply extended to the case of a hard-sphere ensemble in both the Monte Carlo and molecular dynamic approaches. However, the number of particles which may be dealt with for a fixed investment of computer time, the dimension of the fundamental cell and the statistical significance of the computed result impose more severe practical constraints in three dimensions. The results for a hard-sphere assembly are shown in Figures 2.5(a) and 4.6.

The communal entropy of melting arising from the increase in accessible volume of a caged atom on melting is given schematically by the shaded area of Figure 10.3, and represents the difference between the entropy of melting and the entropy if the particles had remained confined to their lattice sites in the solid phase. We see that the communal entropy change represents only a small fraction of the entropy of melting, given by the area of the rectangle shown in Figure 10.3.

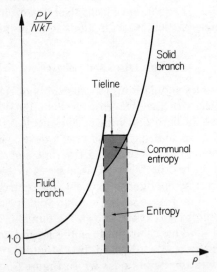

Figure 10.3 Entropy changes in the hard-sphere solid–fluid transition. The total entropy of melting is given by the area of the rectangle, of which the communal entropy shown by the darker shading represents only a small fraction

Lennard–Jones Fluids

Extensive molecular dynamic and Monte Carlo simulations have been reported for 2- and 3-dimensional Lennard–Jones fluids, liquid argon in particular. The principal objectives of the Lennard–Jones computations, in contrast to the essentially theoretical role of the hard-sphere and square-well systems, are to simulate as closely as possible the experimental thermodynamic data, and to provide reference isotherms for the testing of the various approximate theories of the distribution function. In the former case, for non-critical states, discrepancy can in part be attributed to inadequacy of the 12–6 potential, whilst the latter provides a very severe test of the theory—the pair potential being explicitly defined. Nevertheless, the simulation provides a detailed understanding of the atomic structural and kinetic processes which aid the theoretical developments.

Rahman has performed a molecular dynamics simulation on a periodic three-dimensional system of 864 argon atoms interacting through a LJ(6–12) potential ($\varepsilon/k = 119\cdot8$ K, $\sigma = 3\cdot405$ Å, $\rho = 1\cdot374$ g/cm^2, $T = 94\cdot4$ K). The structural features of the system are in good qualitative agreement with the neutron scattering data (Figure 10.4): the radial distribution being determined simply as the distribution of interparticle separations in the assembly. The data is

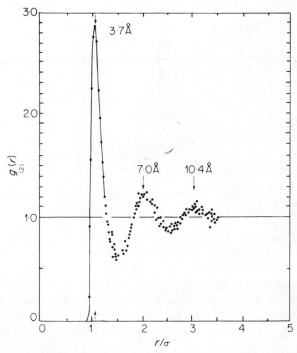

Figure 10.4 Molecular dynamic simulation of the radial distribution function of liquid argon ($\rho = 1\cdot374$ g/cm^2, $T = 94\cdot4$ K)

230

obtained directly from the coordinate distribution and no intermediate transformations of scattering data are necessary. Of more interest, perhaps, are the kinetic features of the system. The molecular dynamics approach allows us to follow in microscopic detail the temporal evolution in phase of a classical system to an extent denied the experimental techniques. Non-equilibrium phenomena can be followed provided the relaxation time is substantially less than the length of the entire chain. Macroscopic consequences of hydrodynamic flow, for example, are therefore excluded from these studies. Steady-state time-dependent phenomena such as self-diffusion and velocity autocorrelation are, however, particularly well suited to the molecular dynamic technique.

The Diffusion Coefficient

If the configurational evolution were a purely random or Brownian process, the probability distribution in the location of a given particle whose initial position was known, would relax as a Gaussian (Figure 10.5). The area beneath

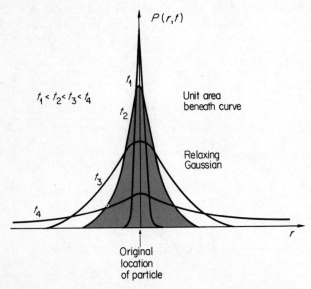

Figure 10.5 Gaussian spatial relaxation in the location of a given particle as a function of time

the curve, normalized to unity, represents the probability of finding the particle *somewhere* in the system. The evolution obviously cannot proceed for such a time that the dimensions of the fundamental cell begin to intervene. All this could be anticipated from our knowledge of the theory of random errors. In consequence, the *mean square displacement* (MSD) is evidently a linear function of time, and any deviation from linearity must represent some departure from a purely random evolution in the diffusive motion of the particle from its original location. Indeed, it is straightforward to show that

$$\langle r^2 \rangle = 6Dt + c \qquad (10.4)$$

where D is the diffusion coefficient and c, a constant, is zero for a random or *Markovian* evolution. In Figure 10.6 we show the mean square atomic displacement obtained in Rahman's molecular dynamics simulation. The mean square

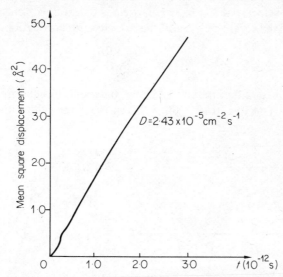

Figure 10.6 Mean square diffusive displacement of argon atom. The curve adopts an asymptotically linear form (equation (10.4)) characteristic of an evolution uncorrelated with its initial history. The short-time evolution shows a departure from linearity associated with initial correlation behaviour. The slope of the linear region yields a diffusion coefficient of $D = 2\cdot43 \times 10^{-5} \, \text{cm}^2 \, \text{s}^{-1}$. Molecular dynamics simulation: $\rho = 1\cdot374 \, \text{g/cm}^2$, $T = 94\cdot4 \, \text{K}$

displacement is a linear function of time after about 2×10^{-12} s, after which its diffusive motions become essentially independent or uncorrelated with its initial history and its subsequent evolution is almost Gaussian. From the asymptotic linear form (10.4) Rahman obtains $D = 2\cdot43 \times 10^{-5} \, \text{cm}^2 \, \text{s}^{-1}$, which is ~ 15 per cent lower than the experimental value at the same density and temperature. Agreement is improved by adjustment of the assumed pair potential. The constant c in equation (10.4) is found to be $0\cdot2 \times 10^{-16} \, \text{cm}^2$, and it tells us whether the diffusion is ahead of or behind the corresponding Markovian process. c being positive, we conclude that diffusion is *impeded* relative to a Markovian evolution. The particles are delayed about their initial positions in their early history and are effectively 'contained' by a relaxing cage of nearest neighbours. At times $< 2 \times 10^{-12}$ s the diffusing particle executes a weak

232

'oscillation' within the cage before 'forgetting' its initial history and is attributed to the persistence in local order. This is most apparent when we consider the velocity autocorrelation and its Fourier transform, the spectral density.

The Velocity Autocorrelation

The velocity of an isolated linear harmonic oscillator is correlated for all time with its initial velocity at $t = 0$. The particle is constrained to execute a linear oscillation of fixed total energy, and the velocity autocorrelation

$$\psi(t) = \frac{\langle \mathbf{v}_i(t) \cdot \mathbf{v}_i(0) \rangle}{\langle \mathbf{v}_i(0)^2 \rangle} \tag{10.5}$$

of an assembly of non-interacting linear harmonic oscillators is shown in Figure 10.7(a). The quantities $\mathbf{v}_i(0)$, $\mathbf{v}_i(t)$ represent the initial and subsequent

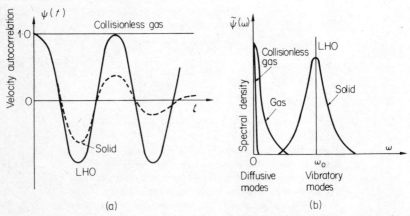

Figure 10.7 (a) Schematic forms of velocity autocorrelation $\psi(t)$ for a collisionless gas, a linear harmonic oscillator and a three-dimensional anharmonic solid. (b) The Fourier transform $\psi(\omega)$ or spectral density of the velocity autocorrelation for the three cases shown in (a). The spectrum is seen to separate into low-frequency diffusive and high-frequency vibratory modes. A liquid would be expected to show both components (cf. Figure 3.7)

velocities, and since for a three-dimensional system the velocity vector can *orientationally decorrelate* with respect to $\mathbf{v}_i(0)$, we take the product $\mathbf{v}_i(0) \cdot \mathbf{v}_i(t)$, and average over the ensemble, $\langle \ \rangle$. The denominator in equation (10.5) normalizes the autocorrelation so that $\psi(t) = 1$. An assembly of interacting three-dimensional oscillators, such as a simple crystalline solid, will show a damped oscillatory autocorrelation and the particles become kinetically decorrelated with their initial velocity coordinates. In the case of a collisionless gas, $\mathbf{v}_i(t) = \mathbf{v}_i(0)$ and $\psi(t) = 1.00$. The decay time of the autocorrelation may be regarded as the phonon lifetime: long phonon lifetimes will appear as peaks in the spectral density distribution, and these are shown in Figure 10.7(b). The spectral densities or frequency spectra are the Fourier transforms of the velocity autocorrelations. We note in particular that in the case of the crystalline solid there are no long-wavelength (low-frequency) contributions: the modes are entirely vibrational. Conversely, in the case of a dilute gas the modes are entirely diffusive, and there are no high-frequency contributions.

In Figure 10.8(a) we compare Rahman's autocorrelation for liquid argon with the Langevin autocorrelation corresponding to a Markovian evolution.

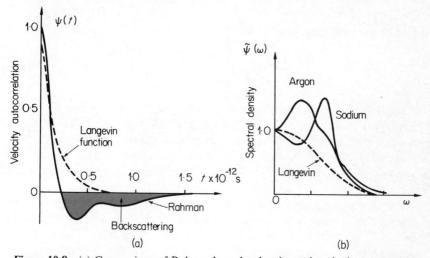

Figure 10.8 (a) Comparison of Rahman's molecular dynamic velocity autocorrelation for liquid argon with the Langevin function appropriate to a totally uncorrelated evolution. The short-time behaviour, before collisional correlations have been established, is seen to be reasonably described by the Langevin function. Thereafter correlation in the form of backscattering (shaded area) induces damped oscillations in the autocorrelation function ($\rho = 1.374 \text{ g/cm}^2$, $T = 94.4 \text{ K}$). (b) Spectral densities of argon, sodium and the Langevin function. The Langevin function is able to describe the low-frequency diffusive component, but cannot account for the high-frequency vibratory modes. This qualitative subdivision is particularly clear in the case of sodium, where the oscillator frequencies are well defined by the nature of the effective pair potential operating in the liquid

The important feature of the liquid-argon function is the negative region corresponding to the backscattering of particles within the relaxing cage of nearest neighbours—dynamically an intermediate case between a dilute gas and an anharmonic solid. Indeed, whilst the particle is making its initial traversal in the caging field its motion is essentially gas-like and the Langevin function provides a reasonable description of its initial evolution. The spectral densities of liquid argon and liquid sodium are shown in Figure 10.8(b); the low-frequency diffusive modes and the high-frequency vibratory modes are clearly apparent in the case of sodium, but not quite so evident in argon. The difference can be largely attributed to the nature of the effective pair potential operating in the fluid—the oscillator frequency is more precisely defined in the case of sodium.

Of particular importance is the non-zero value of $\psi(0)$, indicative of the diffusive motion of the atoms absent in the spectral densities of solids.

The very-long-wavelength ($\omega \sim 0$) modes correspond to purely diffusive motions, and the amplitude $\tilde{\psi}(0)$ is therefore proportional to the coefficient of self-diffusion, D. Since the spectral density $\tilde{\psi}(\omega)$ is the Fourier transform of the velocity autocorrelations $\psi(t)$, we have

$$\tilde{\psi}(\omega) = \int_0^\infty \psi(t) \frac{\sin \omega t}{\omega t}\, dt$$

so that in the long-wavelength limit

$$\boxed{D = \tfrac{1}{3} \int_0^\infty \psi(t)\, dt} \tag{10.6}$$

This important classical result is due to Einstein, and represents the infinite time integral of what is, essentially, the flux of diffusing molecules. It may be shown that the coefficients of viscosity and thermal conductivity can be regarded as essentially diffusion coefficients for momentum and internal energy. This *fluctuation-dissipation* approach to transport phenomena is discussed in greater detail in Chapter 11.

The backscattering shown by the shaded area of the velocity autocorrelation in Figure 10.8(a) obviously lowers the diffusion coefficient, and indeed reduces it to zero in the case of an assembly of harmonic oscillators. The validity of the upper limit in equation (10.6) of $t = \infty$ is assumed here without question: the recurrence of initial phases after an enormously long period ($\sim 10^{145}$ years for a macroscopic system), the Poincaré period, is anticipated on statistical grounds in which case the autocorrelation would regain its initial value. However, setting $t = \infty$ we are implicitly assuming that $t \ll$ Poincaré period.

Quantum Liquids

The simulation methods are not easily applied to quantum-mechanical systems: indeterminism of the particle distributions and conditions on exchange symmetry of the particle wavefunctions are not easily handled in the conventional molecular dynamic and Monte Carlo approaches.

McMillan has calculated the ground-state properties of He^4 by assuming a trial single-particle wavefunction of the form (Figure 10.9)

$$\psi = \exp\left(-a_1/r\right)^{a_2} \tag{10.7}$$

where a_1 and a_2 are variable parameters. Such a wavefunction would be expected to show correlation—that is, the bosons would be excluded from overlapping configurations. The expectation value of the Hamiltonian for spinless bosons acting through a two-body potential is of course

$$\langle \mathcal{H} \rangle = \frac{\int \psi_N \mathcal{H} \psi_N\, d\tau}{\int \psi_N \psi_N\, d\tau} \tag{10.8}$$

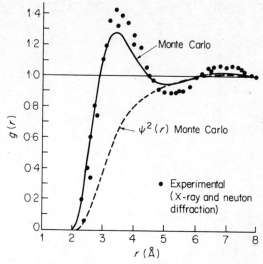

Figure 10.9 Comparison of the Monte Carlo and experimental radial distribution functions of liquid He⁴. The pair correlation $\psi^2(r)$ is also shown

where ψ_N is the N-body product of single-particle functions (10.7) ($N = 32$). The integral (10.8) is evaluated by a Monte Carlo method (cf. equation (10.3)): millions of configurations are generated in a cubical periodic array and any given one is accepted or rejected according to a criterion which ensures that (10.8) converges to the correct thermodynamic limit. The parameters a_1, a_2 are varied so as to minimize $\langle \mathcal{H} \rangle$: taking a pair potential of Lennard–Jones 6–12 form ($\sigma = 2.556$ Å, $\varepsilon/k = 10.22$ K), McMillan finds a minimum ground-state energy at the equilibrium density ($\rho = 2.20 \times 10^{22}$ atoms cm^{-3}) for $a_1 = 2.6$ Å, $a_2 = 5$. The energy at the minimum is $-0.77 \pm 0.09 \times 10^{-15}$ erg atom^{-1} compared with the experimental ground-state energy of -0.988×10^{-15} erg atom^{-1}—a discrepancy of about 20 per cent. At other densities, ranging to beyond the solid–liquid transition at a pressure of 25 atm., a_2 was held fixed at 5 and a_1 adjusted to minimize (10.8) at each density. These values of the variational parameters represent the optimum form for the single-particle trial function, (10.7) (Figure 10.9). The parameters optimized, the configurational (Figure 10.9) and thermodynamic properties follow directly from the weighted phase averages (10.3). In Figure 10.10 we show how the ground-state energy varies with density. There is a significant discrepancy between the experimental and Monte Carlo functions in each case. The trial wavefunction agrees well with the experimental radial distribution regarding the location and sharpness of the cut-off near 2.6 Å; the two variational parameters in the pair function adjust the position and sharpness of the cut-off in the pair function. The experimental curves have a higher peak at 3.5 Å and larger oscillations at large r. This seems to indicate that $\psi(r)$ should peak up somewhat in the region of greatest attraction, but there is no freedom to do this in the two-parameter

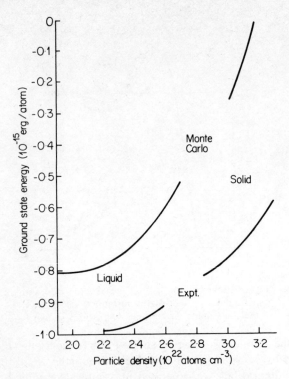

Figure 10.10 The experimental and theoretical ground-state energies of solid and liquid He^4 as functions of density

trial function used by McMillan. The discrepancy in the energy–density curve (Figure 10.10) may similarly be ascribed to shortcomings in the flexibility of the trial function $\psi(r)$.

Surface Phenomena

In Chapter 7 the liquid–vapour density transition profile was established on the basis of statistical mechanical and quasi-thermodynamic analyses. The two routes lead to fundamentally different profiles at temperatures just above the triple point: in the former case the single-particle distribution shows the development of stable density oscillations normal to the liquid surface, whilst the quasi-thermodynamic analysis yields monotonic profiles.

At present there is no *direct* experimental evidence on the basis of which we can decide between the two profiles, although there is a considerable body of indirect evidence which appears to support a structured transition. The machine simulation of a free liquid surface therefore adopts a particularly significant role in aiding our theoretical understanding of the microstructural and micro-kinetic processes in the region of the transition zone.

Croxton and Ferrier solved the one-dimensional BGY equation for the single-particle distribution function across the transition zone for liquid argon at the triple point, and obtained a transition profile showing a weak oscillation (equation (7.16), Figure 7.12). These authors go on to perform a molecular dynamics simulation of the transition zone for a Lennard–Jones system of argon molecules. The greater complexity of surface studies necessitates computation either for fewer atomic centres, resulting in poor statistics, or for a two-dimensional instead of three-dimensional assembly.

Croxton and Ferrier decided upon the latter program as representing the better compromise between expediency and significance of the computed correlation. For a given number of particles a two-dimensional molecular array is, of course, effectively much more extensive than a three-dimensional assembly. The search for periodicity in the surface structure is particularly susceptible to artefacts arising from small replication distances, and this strongly favours the two-dimensional simulation. By making the replication distance large (100 Å) communal entropy defects will be small, and since one of the objectives is to identify the low-entropy surface states, this provision is essential, although it does in practice restrict the computations to two dimensions. Topologically the array is identical to a cylinder of height and circumference 100 Å (Figure 10.11). Elastic reflection of the vapour atoms from the top boundary is perfectly adequate. The lower boundary, however, with its intimate bearing upon the dynamical and configurational coupling of the bulk and surface 'phases' must be considered in more detail. Ideally, we should like to allow a simulated bulk to interact with a simulated surface, whereupon coupling would take care of itself. An expedient embodying the characteristics of the coupling can be used instead, by setting up a matrix of disruptive elements whose magnitudes depend on time and space in a way characteristic of the bulk.

A *static* 'bulk' system of 200 argon atoms is set up below the dynamic array in random positions, and the field of the static assembly is allowed to permeate the dynamic system, indicated by the grey shading in Figure 10.11. The bulk field is now subjected to a sinusoidal space–time modulation having characteristic eigenvalues λ, τ, where λ is the correlation length or range of spatial coherence of a cluster of particles in the liquid (~ 10 Å), while τ is the characteristic lifetime of such a cluster ($\sim 10^{-11}$ s). Such a modulation ensures that on average the dynamic system neither gains nor loses energy, and certainly there is no change in the number of particles.

The dynamic system eventually develops the strongly oscillatory transition profile shown in Figure 10.12. The layered configuration is stable, although only statistically defined: trajectory plots of individual atoms show continuous excursions from one layer to another. More recently Barker and Liu have independently confirmed the oscillatory profile for a three-dimensional system of argon atoms, and Liu has gone on to obtain very satisfactory estimates of the surface tension and its temperature dependence. It does appear that the existence of stable density oscillations at the liquid surface is a real phenomenon and that we should preferentially adopt the oscillatory profile rather than the monotonic.

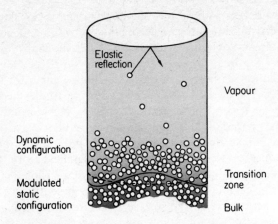

Figure 10.11 The molecular dynamic array in the two-dimensional simulation of Croxton and Ferrier is topologically identical to a cylinder of height and circumference 100 Å. Elastic reflection of the atoms occurs at the upper boundary whilst a thermal disruption matrix at the lower boundary simulates the coupling of the surface and bulk states. The lowest 10 Å of the dynamic array is subjected to the field of a further static array whose force contours are subjected to a space–time modulation thereby simulating the presence of a strongly coupled bulk phase

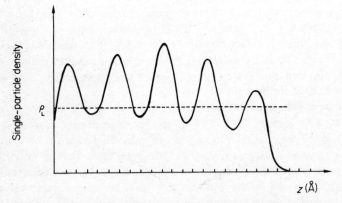

Figure 10.12 The oscillatory density transition profile obtained by Croxton and Ferrier by two-dimensional molecular dynamic simultation at the surface of liquid argon (cf. Figure 7.14)

The pronounced anisotropy at the surface suggests there should be some associated dynamical anisotropy. Croxton and Ferrier have determined the two normalized velocity autocorrelations

$$\psi(t)_{\parallel} = \frac{\langle \dot{x}(0)\dot{x}(t) \rangle}{\langle \dot{x}(0)^2 \rangle} \qquad (10.9a)$$

$$\psi(t)_{\perp} = \frac{\langle \dot{z}(0)\dot{z}(t) \rangle}{\langle \dot{z}(0)^2 \rangle} \qquad (10.9b)$$

and we show $\psi(t)_{\perp}$ in the vicinity of the liquid surface, corresponding to autocorrelation in the normal component of the velocity (Figure 10.13). The pronounced oscillatory nature of this function is immediately evident, and implies

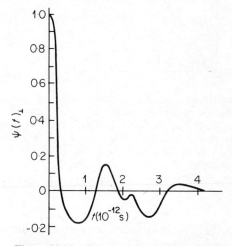

Figure 10.13 The velocity autocorrelation of Croxton and Ferrier normal to the liquid surface. The function shows sustained oscillations consistent with the oscillatory density transition profile

the existence of phonon lifetimes up to 4.5×10^{-12} s (cf. Figure 10.8a). The $\psi(t)_{\parallel}$ autocorrelation is essentially 'liquid-like' and Croxton and Ferrier conclude that the intraplanar dynamic correlations are effectively liquid-like, whilst the interplanar characteristics are those of a more permanent quasi-crystalline system.

The mean square displacement normal and parallel to the surface is shown in Figure 10.14. The gradient of the asymptotically linear region of the curve is proportional to the diffusion coefficient. The asymptotic behaviour does not set in until 3×10^{-12} s due to protracted correlation effects in the initial evolution. The asymptotic value of the two diffusion coefficients is $D_{\perp} = 1.05 \times 10^{-4}$ cm^2 s^{-1} and $D_{\parallel} = 5.22 \times 10^{-5}$ cm^2 s^{-1} at 84 K.

240

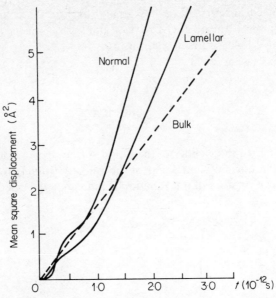

Figure 10.14 The normal and lamellar mean square displacements at the surface of liquid argon at the triple point. The short-time evolution suggests strongly correlated motion, although the linear Markovian evolution is eventually attained. The slope of the linear portion gives the diffusion coefficients

$$D_\perp = 1.05 \times 10^{-4}, \qquad D_\parallel = 5.22 \times 10^{-5}\,\text{cm}^2\,\text{s}^{-1}$$

That $D_\perp$ is larger than $D_\parallel$ is at first sight surprising since we might assume that the quasi-crystalline structure normal to the liquid surface should be characterized by a lower diffusion coefficient more appropriate to a solid. However, on the basis of an atomic hopping or jump diffusion model we should expect the diffusion coefficient normal to the surface to be given by

$$D_\perp = \frac{l^2}{\tau_0} \qquad (10.10)$$

where l^2 is the mean square diffusive step length, and τ_0 is the phonon lifetime during which it defines the lattice site. If we assume interplanar hopping $l \sim 3.8\,\text{Å}$, and from the autocorrelation function we assume a mean phonon lifetime of about $10^{-4}\,\text{cm}^2\,\text{s}^{-1}$, the above relation yields a diffusion constant of about $10^{-4}\,\text{cm}^2\,\text{s}^{-1}$, substantially in agreement with the computed value of $D_\perp = 1.05 \times 10^{-4}\,\text{cm}^2\,\text{s}^{-1}$. The choice of l for a crystalline solid has to be close to the interatomic spacing, $\sim 3.8\,\text{Å}$, but in this case the diffusion coefficient is of the order of $10^{-9}\,\text{cm}^2\,\text{s}^{-1}$, implying a phonon lifetime of about $10^{-7}\,\text{s}$. For lamellar diffusion both l and τ_0 would be smaller, but (10.10) depends more sensitively upon the diffusive step length than τ_0, and undoubtedly the mean lamellar step length will be considerably smaller than 3·8 Å. In consequence $D_\perp > D_\parallel$ at the liquid surface.

Investigation of the Friedel Oscillations

The liquid-metal ion–ion interaction is believed to be fundamentally different from that operating in non-conducting fluids; the distinction is attributed to the presence of conduction electrons. The conduction-electron wavefunctions are orthogonalized with respect to the closed-core states and effectively 'diffract' around the 'opaque' ionic core, setting up *Freidel oscillations* in the screening electron cloud. There is some theoretical support for these oscillations provided the Fermi surface is sharp (Chapter 8). Experimental evidence is not conclusive, however: inversion of scattering data to yield the direct correlation certainly suggests the development of a long-range oscillatory (LRO) interaction (Chapter 4), but, on the other hand, positron annihilation experiments seem to suggest that in some cases the Fermi surface is sufficiently blurred as to damp out the Friedel oscillations (Chapter 8).

Paskin and Rahman have investigated the dynamics and structure of liquid sodium using various LRO potentials by the method of molecular dynamics. Taking each of these potentials in turn they attempt to reproduce the radial distribution function and the coefficient of self-diffusion observed experimentally.

Johnson, Hutchinson and March inverted the sodium neutron scattering data by means of the BGY equation, and obtained the effective potential shown in Figure 10.15(a). Paskin and Rahman used two model potentials; LRO1, which was chosen to fit the experimental potential at the first minimum, and LRO2, which was chosen to give a better fit to the observed RDF. A comparison of the simulation with experiment is shown in Figure 10.15(b). It is seen that LRO1 yields a much sharper $\rho g_{(2)}(r)$ than does the experimental curve. LRO2, however, yields a much more satisfactory RDF, but is quite different from the inversion obtained by Johnson *et al.* in Figure 10.15(a). It does appear that the inversion procedures cannot be depended upon. Again, the molecular dynamics diffusion coefficients on the basis of the two oscillatory potentials appear to bracket the experimental value (Table 10.1).

Table 10.1. Diffusion coefficients for liquid sodium (373 K)

D (expt)	$4\cdot2 \times 10^{-5}$ cm^2 s
D (LRO1)	$1\cdot9 \times 10^{-5}$
D (LRO2)	$5\cdot8 \times 10^{-5}$

Molecular dynamics simulations recently performed by Schiff explicitly isolate the contributions to the structure and velocity autocorrelation due to the Friedel oscillations. Since the amplitude of those oscillations is in all cases expected to be $\ll kT$, the modification of the structure and dynamics is likely

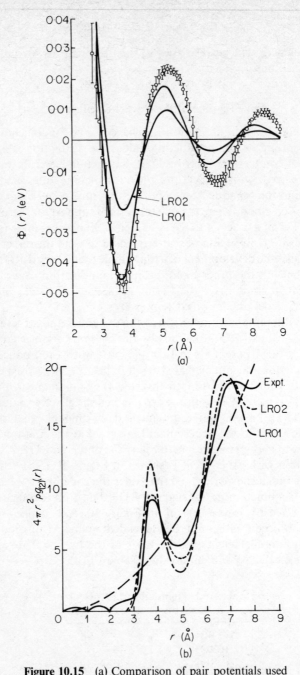

Figure 10.15 (a) Comparison of pair potentials used for liquid sodium by Paskin and Rahman (LRO1, LRO2) with the BGY inversion of experimental data by Johnson, Hutchinson and March. (b) The function $4\pi r^2 \rho g_{(2)}(r)$ for liquid sodium on the basis of experiment and molecular dynamic simulation using the potentials LRO1, LRO2

to be small. Schiff truncates a theoretically realistic sodium ion–ion interaction at $r = 1.53\sigma$ (no oscillations) and $r = 3.20\sigma$ (oscillations) (Figure 10.16). The presence of oscillations has very little effect upon either the structure factor or the velocity autocorrelation.

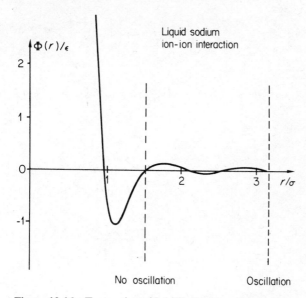

Figure 10.16 Truncation of Schiff's liquid-sodium potential at $r = 1.53\sigma$ (no oscillation) and $r = 3.20\sigma$ (oscillations)

Should the periodicity of the Friedel oscillations coincide with the dominant period of the structure (given by the principal peak $S(k_0)$, in the structure factor) we should expect a 'resonance' in the structure factor. The wavenumber of the Friedel oscillations is $2k_F$ (k_F is the wavenumber at the Fermi surface), and if this is identical to k_0 Schiff points out that such a resonance may occur in a liquid Li–Mg alloy. A resonance occurs at $2k_F = k_0 = 6.8 \text{ Å}^{-1}$ which, on the basis of a free-electron model, should occur around 60 per cent Mg.

Phase Transitions of a Hard-ellipse System

Viellard-Baron has studied by Monte Carlo methods the properties of a system of 170 hard elastic ellipses of axis ratio $a/b = 6$. It is found that the melting transition density from the close-packed configuration (Figure 10.17) depends essentially on the axis ratio a/b: as the ratio increases so does the density at which melting from the solid phase occurs.

In Figure 10.18(a) we show the high-density branch of the isotherm, and a first-order solid–nematic phase transition is observed at a reduced area $A/A_0 = 1.15$, significantly higher than the density at which the hard-disc system starts

244

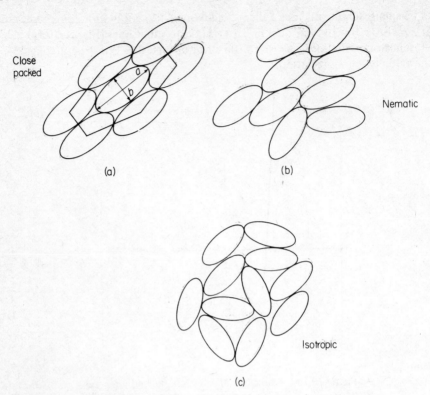

Close packed

Nematic

(a)

(b)

Isotropic

(c)

Figure 10.17   (a) Close-packed hard ellipses. (b) Nematic packing of hard ellipses. (c) Isotropic packing of hard ellipses. The axis ratio $a/b = 6$

to melt. This we may understand in terms of the additional *orientational* degree of freedom available to the system, absent in symmetrical assemblies.

At a reduced area $A/A_0 = 1\cdot40$ a second transition to the nematic phase occurs and is marked by the small van der Waals loop (Figure 10.18b). For small eccentricities this transition occurs in the solid range: for large eccentricities, such as here, it occurs in the fluid range. The oriented fluid is thus analogous to a nematic liquid crystal. The transition is believed to be of first order (i.e. with a discontinuity in the entropy, since the 'unlocking' of the orientation angles leads to a sudden increase in the configurational volume of the system, and hence in the entropy.) This geometrical effect is analogous to the 'unlocking' of the positions which leads to melting in the hard-disc system.

Since the disorientation affects only one degree of freedom per ellipse instead of two in the melting transition, the nematic and isotropic liquid branches of the isotherm are very close to each other. At high densities an order–disorder transition *with respect to the centres of gravity*, analogous to the solid–fluid hard-disc transition, is observed. At a lower density an order–disorder transition *with respect to the orientation of the ellipses* occurs. At high density the major axes

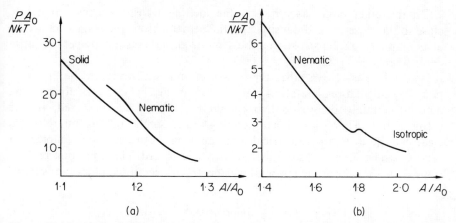

(a) (b)

Figure 10.18 (a) Hard-ellipse Monte Carlo equation of state showing the first-order solid–nematic phase transition at $A/A_0 = 1.15$. (b) First-order nematic–isotropic phase transition at $A/A_0 = 1.40$

have on average, the same orientation so that the Zwetkoff order parameter (Chapter 9)

$$S = \langle \tfrac{3}{2} \cos^2 \theta_i - \tfrac{1}{2} \rangle_N$$

taken as an ensemble average of the molecular orientations θ_i to the mean orientation, yields $S = 1$ for a perfectly ordered system, whilst at lower densities the ellipses can rotate and the orientational order is destroyed, and ultimately $S = 0$ for an isotropic orientationally disordered assembly. The variation of S with packing density is shown in Figure 10.19.

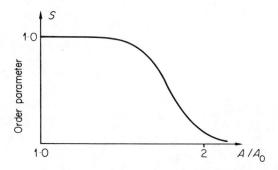

Figure 10.19 Variation of the Zwetkoff order parameter S with packing density of hard ellipses

Viellard-Baron goes on to consider phases of the system for very weak and very strong elongations. If the axis ratio a/b is sufficiently close to 1, then on decreasing the density the system undergoes a disorientation transition before melting—a phase with the ellipses completely disoriented ($S = 0$) but whose centres form a lattice.

On the other hand, as $a/b \to \infty$ the melting transition occurs infinitely close to the close-packed density, and the nematic–isotropic transition occurs at an infinitely low density so that the nematic phase is sustained over the entire density range.

It is quite easy to visualize these two limiting situations. In the first case $(a/b \sim 1)$ orientational rotation can develop at the slightest relaxation of the density, certainly before configurational breakdown has occurred. In the second situation $(a/b \sim \infty)$, corresponding to an assembly of long thin rods, it is clear that there are no geometrical constraints on the distribution of molecular centres, whilst orientational freedom is very restricted. These limiting realizations serve to confirm the essentially geometric aspect of the phase transitions.

References

C. A. Croxton, *LSP*.

B. J. Alder and W. G. Hoover (molecular dynamics) and W. W. Wood (Monte Carlo) in *Physics of Simple Liquids* (Eds. Temperley, Rowlinson and Rushbrooke), North-Holland Publishing Co. (1968), Chapters 4, 5.

CHAPTER 11

Transport Phenomena

Introduction

So far we have been concerned exclusively with *equilibrium* properties of simple liquids—in particular the relation of the structural and thermodynamic features of an assembly to the effective pair potential operating in the system. We now turn our attention to non-equilibrium situations which, unless sustained by some external constraint, will relax irreversibly to the equilibrium configuration. We now ask why it is that this evolution always proceeds irreversibly towards an equilibrium characterized by a maximum entropy (subject to applied constraints), whilst the microscopic equations of motion are necessarily mechanically reversible. This is the very root of the reversibility paradox, and whilst no rigorous results exist at liquid densities we do have some understanding of the irreversible evolution in dilute systems.

If an external constraint is applied, such as a sustained thermal or velocity gradient, there will be associated fluxes of energy, momentum or a diffusive flux in the case of a concentration gradient, and the resulting equilibrium will be a *dynamic* one. These fluxes are characterized by transport coefficients—coefficients of thermal conductivity, shear viscosity and self-diffusion, and rarely develop in isolation—the Joule heat flux associated with electron transport, for example.

Of course, under conditions of non-equilibrium, *distortions* of the isotropic equilibrium distribution $g_{(2)}(r)$ will occur, and these distortions will be rather directly related to the associated flux and therefore to the appropriate transport coefficient. Provided the deviations from equilibrium are not too great, however, the distortions may be treated as perturbations about the isotropic distribution, and so an adequate knowledge of the equilibrium characteristics of the system is a prerequisite to any discussion of transport phenomena.

We have to consider in quite general terms the phase-space evolution of a non-equilibrium assembly towards equilibrium. The discussion must be conducted in phase space since equilibration is characterized by both dynamical and configurational processes, and although the former generally precedes the latter, rapidly becoming Maxwellian so that subsequent equilibration is almost entirely configurational (the Smoluchowski equation), we shall postpone discussion of this situation until later.

We are interested in the N-body phase distribution $f_N(\mathbf{p}^N, \mathbf{q}^N, t)$ defined such that

$$f_{(N)}(\mathbf{p}^N, \mathbf{q}^N, t) \, d\mathbf{p}^N \, d\mathbf{q}^N$$

represents the number of particles whose representative points lie within the

$6N$-dimensional phase element $dp^N dq^N$ located at $(\mathbf{p}^N, \mathbf{q}^N)$. This function is normalized such that

$$\iint f_{(N)}(\mathbf{p}^N, \mathbf{q}^N, t)\, d\mathbf{p}^N\, d\mathbf{q}^N = N \tag{11.1}$$

If the molecules are uniformly distributed in space we also have

$$\int f_{(N)}(\mathbf{p}^N, \mathbf{q}^N, t)\, d\mathbf{p}^N = \frac{N}{V^N} \tag{11.2}$$

The Boltzmann Transport Equation

The aim of kinetic theory is to describe the phase evolution of a non-equilibrium distribution $f_{(N)}(\mathbf{p}^N, \mathbf{q}^N, t)$ towards equilibrium, and we enquire how the population of the element $(d\mathbf{p}^N, d\mathbf{q}^N)$ varies as a function of time. If there are no molecular collisions then the cloud of representative points in phase space drifts through the phase elements such that a molecule with coordinates $(\mathbf{p}, \mathbf{q})$ at time t will have the coordinates

$$\left(\mathbf{q} + \frac{\mathbf{p}}{m}\delta t, \frac{\mathbf{p}}{m} + \frac{\mathbf{X}}{m}\delta t \right)$$

at time $t + \delta t$, where $\mathbf{X}$ is an external force acting on the molecule. We therefore have

$$f\left(\mathbf{q} + \frac{\mathbf{p}}{m}\delta t, \frac{\mathbf{p}}{m} + \frac{\mathbf{X}}{m}\delta t \right) = f(\mathbf{p}, \mathbf{q}, t) \tag{11.3}$$

The flux of representative points through phase space behaves as if it were an incompressible fluid, as indeed it must if it is to bear the interpretation of a flux of probability density.

If there are collision the above expression must be modified to account for collisional population and depopulation of the phase element:

$$f\left(\mathbf{q} + \frac{\mathbf{p}}{m}\delta t, \frac{\mathbf{p}}{m} + \frac{\mathbf{X}}{m}\delta t \right) = f(\mathbf{p}, \mathbf{q}, t) + \left(\frac{\partial f}{\partial t} \right)_{coll} \delta t \tag{11.4}$$

The term $(\partial f/\partial t)_{coll}$ is generally designated the *collision integral*, J, and it specification allows us to describe the equilibration process in terms of impulsive binary encounters.

Expanding the left-hand side of (11.4) to first order in δt, we obtain the evolution of the distribution $f_{(N)}(\mathbf{p}_N, \mathbf{q}_N, t)$ as we let $\delta t \to 0$. This is the Boltzmann transport equation

$$\frac{\partial f_{(N)}}{\partial t} + \sum_{i=1}^{N} \left\{ \frac{\mathbf{p}_i}{m} \cdot \frac{\partial f_{(N)}}{\partial \mathbf{q}_i} + \mathbf{X}_i \cdot \frac{\partial f_{(N)}}{\partial \mathbf{p}_i} \right\} = \left(\frac{\partial f_{(N)}}{\partial t} \right)_{coll} \tag{11.5}$$

For present purposes we do not need to develop explicit expressions for the collision integral J here—for these the reader is directed to *LSP*. Nonetheless

Boltzmann determined a form for the collision integral on the basis of following assumptions:

(i) The gas is assumed sufficiently dilute that only *binary* encounters need be considered.

(ii) There are no 'persistence of velocity' effects after a collision. This is often known as the 'molecular-chaos' assumption.

(iii) The scattering process is assumed independent of the external field **X**.

(iv) The distribution $f_{(1)}(\mathbf{p}, \mathbf{q}, t)$ does not vary appreciably during a time interval of the order of a molecular collision, nor does it vary appreciably over a spatial distance of the order of the range of the intermolecular forces. This restricts the description to large momentum exchange *impulsive* binary encounters.

It has been shown that subject to these assumptions the Boltzmann transport equation describes an *irreversible* evolution in phase, even though the microscopic motions are dynamically reversible. We shall not pursue this aspect of transport further here, except to say that the irreversibility may be directly attributed to the assumption of molecular chaos according to which the system 'forgets' its preceding history and, as it were, frees the dynamical behaviour of the molecules from the reversibility implied in their equation of motion.

Obviously the assumption of molecular chaos is inappropriate at liquid densities when there is extensive spatial and dynamical correlation within the assembly, and for realistic systems, when the particles are in extended interaction. The aspect of irreversibility—the Boltzmann H-theorem and the approach to equilibrium—will be discussed further in the next chapter.

The Fokker–Planck Equation

In high-density, low-temperature systems of particles in continuous interaction the concept of a binary encounter is no longer meaningful and the molecular-chaos hypothesis can no longer be applied to ensure an irreversible evolution in phase. Under these circumstances of 'soft' coupling the momentum exchanges are generally small such that $\mathbf{p} \gg \Delta\mathbf{p}$: the large-momentum binary exchanges discussed above no longer govern the evolution. Whilst in the case of binary impulsive encounters the colliding centres execute dynamically complementary conservative trajectories, a fundamental assymmetry arises in the case of the high-density, extended-interaction system, for whilst the dynamical interaction between a particle and its neighbours is collectively both reciprocal and conservative, the net effect on the neighbourhood is insignificant.

The differential equation describing the single-particle evolution in phase in the weak coupling limit ($\Delta\mathbf{p} \ll \mathbf{p}$) is the Fokker–Planck equation

$$\frac{\partial f_{(1)}}{\partial t} + \frac{\mathbf{p}_1}{m}\frac{\partial f_{(1)}}{\partial \mathbf{q}_1} + \mathbf{X}_1\frac{\partial f_{(1)}}{\partial \mathbf{p}_1} = M \qquad (11.6)$$

where M is the 'soft' equivalent of the 'hard' collision operator J appearing in the Boltzmann transport equation. We shall not need to solve the Fokker–Planck equation in general since the momentum evolution very rapidly attains Maxwell–Boltzmann form, and its substitution in equation (11.6) provides a purely configurational projection of the Fokker–Planck equation. Thus, provided we don't want information about the very early stages of the total evolution, its configurational projection, the *Smoluchowski equation*, will tell us all we need to know about the configurational approach to equilibrium. Of course, we need to know that the momentum equilibration is complete, and whilst the molecular dynamics simulations seem to support the suggestion that dynamic equilibration is rapidly completed, the *rate* at which dynamic equilibrium is approached is governed by the *friction constant* about which we shall have more to say later.

The sort of process we might envisage for the application of the Fokker–Planck equation, or its configurational projection, the Smoluchowski equation, is Brownian motion in which a microscopic particle executes a random trajectory in the fluctuating soft force field of its neighbours. We shall return to this when we come to discuss the Smoluchowski equation.

We cannot of course apply the molecular-chaos hypothesis to decouple the deterministic reversible evolution into an indeterministic irreversible one. Kirkwood observed that the force fluctuations on a representative molecule remain correlated only for a short period of time, τ, after which they become essentially statistically independent. If we apply a coarse-grained time-smoothing over a period τ, then adjacent evolutionary events will be statistically independent. This, of course, is a natural extension of the hypothesis of molecular chaos. Kirkwood's great contribution was to establish a direct connection between the friction constant which governs the rate of momentum equilibration, and the lifetime τ of the force autocorrelation (equation (11.13)): this we shall discuss shortly.

Realistic Interactions—the Rice–Allnatt Equations

In a real fluid we imagine the molecular motion to consist of infrequent impulsive binary encounters followed by a Brownian motion in the fluctuating force field of the environment: this is clear from the representative molecular dynamics trajectory shown in Figure 3.7(a). Thus far we have discussed the single-particle phase evolution in the two extreme limits of large-momentum binary exchanges appropriate to impulsive encounters between rigid repulsive particles, and in the small-momentum exchanges for a system of particles in extended collective interaction. In both cases the mechanically reversible Liouville equation was found to develop characteristics of irreversibility in establishing the collision integral J (through the assumption of molecular chaos) in the Boltzmann transport equation, and the Fokker–Planck analogue M (through the friction constant and time smoothing procedures).

Rice and Allnatt have suggested that a more appropriate transport equation at liquid densities would be

$$\frac{\partial f_{(1)}}{\partial t} + \frac{\mathbf{p}_1}{m}\frac{\partial f_{(1)}}{\partial \mathbf{q}_1} + \mathbf{X}_1\frac{\partial f_{(1)}}{\partial \mathbf{p}_1} = J + M \tag{11.7}$$

combining the J and M processes. It is assumed that the processes J and M operate independently.

This approach implicitly assumes an interaction potential of the form shown in Figure 11.1, although Rice and Allnatt maintain that extension to more realistic interactions presents no formal difficulty.

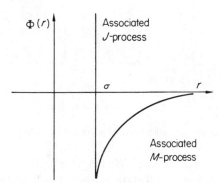

Figure 11.1 Model pair interaction used by Rice and Allnatt showing regions associated with J and M processes

Associated with the J and M processes will, of course, be a 'hard' and 'soft' friction coefficient, β_H and β_S, respectively. It is the determination of the friction constants which at present limits the application of the transport equations to largely formal solutions.

The Smoluchowski Equation

We have pointed out that for dense systems at low temperature, equilibrium in momentum space tends to precede configurational equilibration. In a dense fluid it is primarily the dissipative quasi-Brownian motion (M-coupling) in the soft force field that is responsible for the momentum evolution, whilst it is the strong hard-core J-coupling which governs the geometrical process of configurational evolution. Associated with these processes are the soft and hard friction constants discussed above.

If $\beta_S \gg \beta_H$ we may assume that momentum evolution is essentially complete, before the spatial coordinates in fact have had time to change significantly from their initial values so that further equilibration occurs in configuration space alone. In this case the real-space projection of the general Fokker–Planck phase evolution is the Smoluchowski equation, and this equation enables us to discuss non-equilibrium situations after the initial dynamic transients have died down.

The classical theory of Brownian motion as developed by Einstein in 1905 provides an approach to the Smoluchowski equation with application to dense soft-coupled fluids. The configurational diffusion of particles in a fluid is shown schematically in Figure 11.2, and we may interpret the process at a

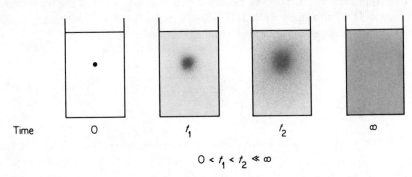

Time   0 t_1 t_2 ∞

$$0 < t_1 < t_2 \ll \infty$$

Figure 11.2 Diffusive relaxation of a concentration gradient with time

macroscopic or a microscopic level. We may, for example, regard the equilibration as a relaxation of a concentration gradient c with time, or alternatively we may imagine the individual particles to execute random Brownian trajectories until they have spread uniformly throughout the accessible volume.

In the macroscopic case the evolution is determined by the diffusion equation

$$\frac{\partial c}{\partial t} = D\nabla^2 c \tag{11.8}$$

where D is the coefficient of self-diffusion, and it is straightforward to show that $D = kT/\beta_{(1)}$. The result is important in that it describes an irreversible configurational evolution towards equilibrium in the fluctuating soft force field. The natural extension of this equation from our point of view is towards the molecular interpretation.

Smoluchowski showed that the corresponding generalization of (11.8) in an external field $\mathbf{X}$, with c replaced by $n_{(1)}$, is

$$\frac{\partial n_{(1)}}{\partial t} = \mathrm{div}\left\{ \frac{kT}{\beta_{(1)}}\,\mathrm{grad}\,n_{(1)} - \frac{\mathbf{X}}{\beta_{(1)}}n_{(1)} \right\} \tag{11.9}$$

in vector notation. This is *Smoluchowski's equation*. (Setting $\mathbf{X} = 0$, (11.9) reduces to the original diffusion equation.) Clearly, equilibrium is characterized by the absence of generalized fluxes in the system ($\partial n_{(1)}/\partial t = \mathrm{div}\{\ \} = 0$)—but then this is intuitively obvious. Equation (11.9) could, of course, have been obtained from the Fokker–Planck equation by insertion of the Maxwell–Boltzmann velocity distribution.

There is one important modification which we have to make to this treatment before it can be used to describe the configurational evolution in a dense fluid. The Brownian particles are regarded as being *independent*, as in a dilute gas, so that all the structural information is contained in the single-particle distribution $n_{(1)}$. In a dense fluid assembly we know that there are well-developed spatial correlations, and we really need to know how the hierachy of higher-order distributions evolve, although in the pair approximation we need only consider the evolution of the two-particle function, $n_{(2)}$.

The necessary modifications of equation (11.9) have been made by Kirkwood and by Eisenschitz. Now we consider the representative *pair* of molecules as the Brownian particle moving in the fluctuating soft force field of the remaining $(n-2)$ particles. The irreversible progress towards equilibrium is now expressed in terms of a *two-body friction constant* $\beta_{(2)}$ given by the force autocorrelation acting on the representative *pair*.

$$\beta_{(2)} = \frac{1}{3mkT} \int_0^T \langle F_{(2)}(0) \cdot F_{(2)}(t) \rangle_{(N-2)} \, dt \qquad (11.10)$$

The upper limit is again set at τ rather than ∞ to avoid Poincaré recurrence difficulties, and irreversibility is ensured by time smoothing over the period of decay of the force autocorrelation, τ. Assuming $\beta_{(2)}$ is large enough we regain the configurational projection of the two-body Fokker–Planck equation

$$\frac{\partial n_{(2)}}{\partial t} = \text{div} \left\{ \frac{kT}{\beta_{(2)}} \text{grad } n_{(2)} - \left(-\frac{\text{grad } \Psi}{\beta_{(2)}} \right) n_{(2)} \right\} \qquad (11.11)$$

which is the two-body Smoluchowski equation. Instead of the external force acting on a Brownian particle we have the mean relative force of one molecule on the other, $-\text{grad } \Psi(12)$, where $\Psi(12)$ is the potential of mean force. At equilibrium $\partial n_{(2)}/\partial t = 0$ and the generalized current $\mathscr{J}_{(2)} = \text{div} \{ \ \}$ is zero; i.e.

$$\frac{kT}{\beta_{(2)}} \text{grad } n_{(2)} = -n_{(2)} \frac{\text{grad } \Psi}{\beta_{(2)}} \qquad (11.12)$$

which follows directly from the definition of the potential of mean force $n_{(2)} = \exp(-\Psi/kT)$.

All this requires a knowledge of the two-body friction constant $\beta_{(2)}$: this, not surprisingly, is theoretically even more elusive than $\beta_{(1)}$.

The Friction Constant

Before we are able to solve the Fokker–Planck, Rice–Allnatt or Smoluchowski equations for the single-particle evolution, we have first to determine the friction constants which govern the rate of the equilibration process.

254

We have seen that the friction constant is formally given as the time integral of the force autocorrelation on particle 1:

$$\beta_{(1)} = \frac{1}{3mkT} \int_0^\tau \langle \mathbf{F}_{(1)}(0) . \mathbf{F}_{(1)}(t) \rangle_{(N-1)} \, dt \qquad (11.13)$$

and if the single-particle friction constant is to be independent of τ, then the integrand must decay to zero. (This does not mean, incidentally, that we may therefore set the upper limit to ∞ : τ must be long with respect to the dynamical relaxation but short with respect to the Poincaré period when initial phases may be recovered). The physical significance of (11.13) is straightforward. The force autocorrelation Φ is likely to have the qualitative form shown in Figure 11.3(a), and the spectral distribution in Figure 11.3(b). The spectral distribution

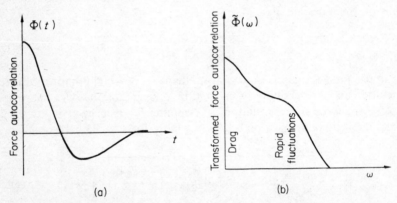

Figure 11.3 (a) Schematic form of force autocorrelation. (b) Spectral density (Fourier transform) of force autocorrelation. The low-frequency component represents a non-fluctuating drag term

is seen to be qualitatively subdivided into a rapidly fluctuating component which, over a period τ, has no net effect, and a low-frequency or *drag* (friction) term (Figure 11.4). The friction coefficient is therefore determined as the long-wavelength hydrodynamic limit

$$\underset{\omega \to 0}{\text{Lt}} \int_0^\tau \tilde{\Phi}(\omega) \, e^{i\omega t} \, dt \qquad (11.14)$$

That is, $\beta_{(1)}$ is proportional to the hydrodynamic drag amplitude $\tilde{\Phi}(0)$. Because of the reversible nature of the underlying particle dynamics τ must be smaller than the Poincaré cycle. If it were not, the physical system would not show irreversible behaviour, and $\beta_{(1)}$ would have the value zero when averaged over the complete cycle.

The evaluation of the time integral of the force autocorrelation presents a formidable mathematical task, and early optimism that reasonably accurate estimates of $\beta_{(1)}$ would be obtainable has now given way to approximate order-of-magnitude determinations.

Instantaneous force-field
distributions

Figure 11.4 Instantaneous force-field
distributions acting on a stationary and
a moving particle

Kirkwood *et al.* suggested writing the single-particle force autocorrelation in the form

$$\langle \mathbf{F}_{(1)}(0) \cdot \mathbf{F}_{(1)}(t) \rangle_{(N-1)} = \langle \mathbf{F}_{(1)}(0)^2 \rangle_{(N-1)} \psi(t) \tag{11.15}$$

where $\psi(s)$ is some monotonic decreasing function of time chosen such that $\psi(0) = 1\cdot 0$. We might propose that $\psi(t)$ decays as the average velocity of particle 1:

$$d\mathbf{v}_{(1)}(t) = -\frac{\beta_{(1)}}{m} \mathbf{v}_{(1)} \, dt$$

i.e.

$$\psi(t) \propto \exp\left(-\frac{\beta_{(1)}t}{m}\right) \tag{11.16}$$

so that we are able to replace the time integral in (11.13) by $m/\beta_{(1)}$. If the environment is in local equilibrium, then the mean square force on particle 1 will be, for purely radial interactions,

$$\frac{\langle \mathbf{F}_{(1)}(0)^2 \rangle}{kT} = 4\pi\rho \int_0^\infty [\nabla\Phi(r)]^2 g_{(2)}(r) r^2 \, dr \tag{11.17}$$

The final expression for the friction constant then becomes

$$\beta_{(1)}^2 = \frac{4\pi\rho}{3} \int_0^\infty [\nabla\Phi(r)]^2 g_{(2)}(r) r^2 \, dr \tag{11.18}$$

Direct experimental measurement of $\beta_{(1)}$ is not possible. We may, however, relate the coefficient of self-diffusion D to the friction constant through the Einstein mobility relation

$$D = \frac{kT}{m\beta_{(1)}} \tag{11.19}$$

A comparison of the experimental and theoretical diffusion coefficients for liquid argon on the basis of equation (11.18) is shown in Table 11.1.

Table 11.1. Diffusion coefficients for liquid argon

| | $D \times 10^5 \, cm^2 \, s^{-1}$ | | |
	84 K	90 K	100 K
Experiment	1·84	2·35	3·45
Equation (11.18)	2·25	2·49	2·92

It has been shown that the assumption made for the form of $\psi(t)$ in (11.15) implies a Gaussian decay of the momentum autocorrelation, and from what we know of the theory of random errors, this kind of evolution is what we would expect from a completely random and uncorrelated velocity autocorrelation. We see from Figure 11.6 that a Gaussian decay gives a quite inadequate representation of $\psi(t)$. The neglect of the negative region of the autocorrelation in a Gaussian model leads to an overestimate of the diffusion coefficient; this we anticipated in equation (11.18).

The Rice–Allnatt model, involving a 'hard' and 'soft' friction constant $\beta_{(1)} = \beta_H + \beta_S$, is able to incorporate both the J and M processes corresponding to the occasional large momentum binary exchanges interspersed with the rapidly fluctuating small-momentum-exchange diffusive motions. Using only the hard-core component of the pair potential in equation (11.18), Rice and Allnatt obtain

$$\beta_H = \tfrac{3}{8}\rho\sigma^3 g_{(2)}(\sigma)(\pi m k T)^{\frac{1}{2}} \tag{11.20}$$

where σ is the atomic diameter. The value of the soft friction coefficient may then be deduced from experimental measurements of D, so that

$$\beta_S = \frac{kT}{mD} - \beta_H \tag{11.21}$$

where β_H is given in (11.20). The values of β_S and β_H are given in Table 11.2, again for liquid argon. In the Rice–Allnatt model, the basic assumption is that

Table 11.2. Friction constants for liquid argon

T (K)	$\rho \, (g \, cm^{-3})$	$(kT/D) \times 10^{10}$ $(g \, s^{-1})$	$\beta_H \times 10^{10}$ $(g \, s^{-1})$	$\beta \times 10^{10}$ $(g \, s^{-1})$
90	1·38	5·11	0·64	4·47
128	1·12	2·94	0·94	2·00
133·5	1·12	3·13	1·00	2·13
185·5	1·12	3·20	1·52	1·68

the Brownian motion induces a sufficiently rapid relaxation to equilibrium of the momentum of the molecule that successive hard-core collisions may be regarded as almost independent. The soft friction coefficient is seen from

Table 11.2 to be many times more efficient in bringing about equilibrium. The importance of the soft component diminishes with increasing temperature such that when $T \gg \varepsilon/k$,

$$\beta_H \gg \beta_S$$

In this limit the weak coupling Fokker–Planck operator M becomes of secondary importance in comparison with the hard-core friction coefficient, β_H.

Viscous Shear Flow

For an interacting system of particles there will be both kinetic and potential contributions to the coefficient of shear viscosity:

$$\eta = \eta_K + \eta_V \tag{11.22}$$

From elementary kinetic theory we have for a dilute gas $\eta_K \propto \rho \bar{c} \lambda$, where $\bar{c}$ is the mean atomic velocity and λ is the mean free path. For a dense fluid, however, η_V dominates and will obviously depend upon the perturbed structure of the assembly. Since we imagine a steady shear to be sustained, the equilibrium will be a dynamic one, and we may use the Smoluchowski equation to relate η_V to the pair distribution which we assume will distort in response to the applied velocity gradient, adopting the time-independent form

$$n_{(2)}(\mathbf{r}) = \rho_0 g^0_{(2)}(r)[1 + w(\mathbf{r})] \tag{11.23}$$

$w(\mathbf{r})$ represents the distortion of the equilibrium distribution, and will be determined by means of the Smoluchowski equation. For the specification of the viscous flow we let the fluid velocity $\mathbf{c}$ depend upon coordinates in an arbitrary way, subject to the condition of incompressible flow, div $\mathbf{c} = 0$. The bulk flow of the liquid may be regarded as exerting a differential dragging effect in each of the molecules of the pair, given by $m\beta_{(2)}\mathbf{c}$, and therefore supplements the mean force term. Neglecting terms of magnitude cw, since both are assumed small, the steady flow is then given by direct substitution into the Smoluchowski equation:

$$
\begin{aligned}
\mathscr{I}_2 = {} & \left\{ \frac{kT}{m\beta_{(2)}} \operatorname{grad} \left(\rho_0 g^0_{(2)}(r)[1 + w(\mathbf{r})] \right) \right. \\
& \left. + \frac{1}{m\beta_{(2)}} (\operatorname{grad} \Psi - m\beta_{(2)}\mathbf{c}) \rho_0 g^0_{(2)}(r)[1 + w(\mathbf{r})] \right\} \\
= {} & \left\{ -[1 + w(\mathbf{r})] \frac{kT}{m\beta_{(2)}} \rho_0 \cdot \langle \mathbf{F}_{(2)} \rangle + \rho_0 g^0_{(2)}(r) \frac{kT}{m\beta_{(2)}} \operatorname{grad} w(\mathbf{r}) \right. \\
& \left. + \frac{kT}{m\beta_{(2)}} \rho_0 \langle \mathbf{F}_{(2)} \rangle - \frac{kT}{m\beta_{(2)}} \rho_0 \langle \mathbf{F}_{(2)} \rangle \cdot w(\mathbf{r}) - c\rho_0 g^0_{(2)}(r) \right\} \tag{11.24}
\end{aligned}
$$

258

where

$$\operatorname{grad} g_{(2)}^0(r) = -g_{(2)}^0(r) \operatorname{grad}\left(-\frac{\Psi}{kT}\right) = +\frac{\langle \mathbf{F}_{(2)}\rangle}{kT} g_{(2)}^0(r)$$

i.e.

$$\mathscr{J}_2 = \rho_0 g_{(2)}^0(r)\left\{\frac{kT}{m\beta_{(2)}} \operatorname{grad} w(\mathbf{r}) - \mathbf{c}\right\} \tag{11.25}$$

From the identity

$$\operatorname{div}(a\mathbf{A}) = \mathbf{A} \cdot \operatorname{grad} a + a \operatorname{div} \mathbf{A} \qquad \text{and} \qquad \operatorname{div} \mathbf{c} = 0$$

the function $w(\mathbf{r})$ is determined from the time-independent Smoluchowski equation

$$0 = \operatorname{div} \mathscr{J}_2$$

i.e.

$$\nabla^2 w(\mathbf{r}) - \frac{\langle \mathbf{F}_{(2)}\rangle}{kT} \operatorname{grad} w(\mathbf{r}) = -\frac{\beta_{(2)}m}{kT}\mathbf{c} \cdot \langle \mathbf{F}_{(2)}\rangle \tag{11.26}$$

subject to the boundary condition

$$\operatorname{grad} w(\mathbf{r}) = 0 \qquad \text{at } r = \infty$$

so that the flow at infinity is, from (11.25), $\rho_0 g_{(2)}^0(r)\mathbf{c}$.

For the purpose of calculation, and without loss of generality, we assume the flow is laminar and Newtonian with a rate of shear $\frac{1}{2}\alpha$ having Cartesian velocity components $c_x = c_z = 0$; $c_y = \alpha x$ (Figure 11.5a). The right-hand side

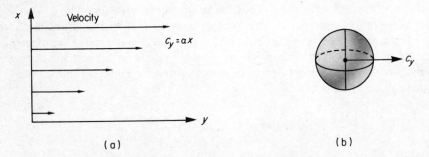

Figure 11.5 (a) Linear velocity gradient for the determination of shear viscosity. (b) The spherical harmonic $P_2^2(\cos\theta)$

of (11.26) then becomes

$$-\frac{\alpha\beta_{(2)}m}{(kT)^2}\frac{\partial\Psi}{\partial r} \cdot r \cdot \sin^2\theta \sin\phi \cos\phi \tag{11.27}$$

Assuming as we must, the same angular dependence on the left-hand side we have

$$w(\mathbf{r}) = \frac{\alpha \beta_{(2)} m}{(kT)^2} u(r) \sin^2 \theta \sin \phi \cos \phi = \frac{\alpha \beta_{(2)} m}{(kT)^2} u(r) P_2^2(\cos \theta) \qquad (11.28)$$

i.e.

$$n_{(2)}(\mathbf{r}) = \rho_0 g_{(2)}^0(r) \left[P_0^0(\cos \theta) + \frac{\alpha \beta_{(2)} m}{(kT)^2} \cdot u(r) P_2^2(\cos \theta) \right] \qquad (11.29)$$

(see Figure 11.5b), where $u(r)$ is an as yet unknown function of the separation and represents the purely *radial* distortion of the equilibrium pair distribution, P_0^0, P_2^2 are the zeroth- and second-order associated Legendre functions. Substitution of (11.28) into (11.26) yields the second-order differential equation describing the radial distortion $u(r)$:

$$\frac{d^2 u}{dr^2} + \left(\frac{2}{r} - \frac{1}{kT} \frac{d\Psi}{dr} \right) \frac{du}{dr} - \frac{6}{r^2} = -\frac{m \beta_{(2)} \alpha}{(kT)^2} \frac{d\Psi}{dr} r \qquad (11.30)$$

Presuming a knowledge of the two-body friction constant $\beta_{(2)}$, solution of the above differential equation for $u(r)$ and substitution in

$$\eta_V = \frac{2\pi \rho_0^2}{15} \beta_{(2)} \int_0^\infty \frac{d\Phi(r)}{dr} g_{(2)}^0(r) u(r) r^3 \, dr \qquad (11.31)$$

should yield an estimate of the coefficient of shear viscosity.

The solution of equation (11.30) has been the centre of some controversy. Being of second order it requires two boundary conditions for its solution, and there is general agreement on one of these:

$$u(r) \to 0 \quad \text{as } r \to \infty$$

For the second condition, Kirkwood, Buff and Green take the 'weak' condition

$$u(r) \to 0 \quad \text{as } r \to 0$$

Eisenschitz objects to this condition on the grounds that it leads to a degree of anisotropy not observed experimentally, although some recent careful work of Champion seems to suggest that Eisenschitz's alternative 'strong' condition on $u(r)$, again at infinity,

$$r^3 u(r) \to 0 \quad \text{as } r \to \infty$$

cannot be accepted without reservation. The Eisenschitz radial distortion function becomes infinite at the origin, but the calculated viscosity may still remain finite since $g_{(2)}^0 \to 0$ as $r \to 0$. We shall not discuss this problem further here except to say that numerical evaluation of (11.31) will not resolve the dilemma because of our inadequate knowledge of $\beta_{(2)}$ at present. For further details of this, and the application of the Smoluchowski equation to the problem of thermal conductivity, the reader is referred to *LSP*.

The Velocity Autocorrelation

From our discussion of the single-particle friction constant in the last section it was apparent that the assumption of a Gaussian decorrelation of the momentum did not provide an adequate representation of the kinetic processes operating in a dense liquid. Such a decorrelation would be appropriate to a random evolution which, from the molecular dynamics simulations, is known not to occur. The spatial and dynamical coupling between a system of particles at high density can only serve to establish a correlation between adjacent dynamical events: only for a system so dilute that negligible backscattering of a particle occurs within its cage of neighbours can we anticipate purely Gaussian behaviour. Otherwise, we must expect an initially Gaussian decay of the velocity autocorrelation during which time the particle executes a quasi-Brownian motion in the fluctuating force field of the cage interior. Then a large momentum exchange with the confining cage will occur, and the representative particle will suffer a velocity reversal. It is in this period that the departure from Gaussian behaviour is most pronounced. We saw in Rahman's molecular dynamics simulation of liquid argon (Chapter 10) that for short times $\sim \bar{d}/\sqrt{T}$, where $\bar{d}$ is the mean atomic spacing, the evolution was essentially Gaussian. Thereafter the development of non-Gaussian terms, 10–20 per cent of the Gaussian term, was observed. After about 10^{-11} s the Gaussian form was recovered. This result is entirely in agreement with the qualitative picture presented above.

In the absence of external forces, the equation of motion of a representative particle may be described as a diffusing Einstein oscillator:

$$m\ddot{\mathbf{x}}(t) = -\beta\dot{\mathbf{x}}(t) - \gamma\mathbf{x}(t) + \mathbf{F}'(t) \tag{11.32}$$

where $\beta\dot{\mathbf{x}}(t)$ represents the frictional drag term, $-\gamma\mathbf{x}(t)$ represents the harmonic restoring term, and $\mathbf{F}'(t)$ represents the rapidly fluctuating force due to the motion of the neighbouring particles. Expressing (11.32) in terms of the particle momentum $\mathbf{p}$ and taking the time derivative and ensemble average of both sides, we have

$$\langle\ddot{\mathbf{p}}\rangle + \beta\langle\dot{\mathbf{p}}\rangle + \gamma\langle\mathbf{p}\rangle = 0 \tag{11.33}$$

The term $\langle\dot{\mathbf{F}}'\rangle$ always vanishes, of course. It now remains to establish a relation between $\langle\mathbf{p}\rangle$ and $\psi(t)$, the velocity autocorrelation. The normalized autocorrelation is related to the conditionally averaged momentum by

$$\psi(t) = \frac{\langle\mathbf{p}(0)\cdot\mathbf{p}(t)\rangle}{\langle\mathbf{p}(0)\mathbf{p}(0)\rangle} = \frac{\langle\mathbf{p}(0)\cdot\mathbf{p}(t)\rangle}{mkT} \tag{11.34}$$

whereupon we have

$$\ddot{\psi} + \beta\dot{\psi} + \gamma\psi = 0 \tag{11.35}$$

which is to be solved for the velocity autocorrelation $\psi(t)$ subject to the boundary conditions $\psi(0) = 1$, $\dot{\psi}(0) = 0$.

The solution is

$$\psi(t) = \frac{1}{\theta_+ - \theta_-}\{\theta_+ \exp(\theta_- t) - \theta_- \exp(\theta_+ t)\} \qquad (11.36)$$

where θ_+ and θ_- are the roots of the auxiliary equation

$$\theta_\mp = -\frac{\beta}{2} \mp \left(\frac{\beta^2}{4} - \gamma\right)^{\frac{1}{2}} \qquad (11.37)$$

When $\beta^2 > 4\gamma$ the autocorrelation $\psi(t)$ is a monotonic function of time representing a Brownian evolution in which there are no restoring terms, whilst if $\beta^2 < 4\gamma$ it is a damped oscillatory function. This model is in qualitative agreement with the molecular dynamics results of Rahman (Figure 11.6). We may

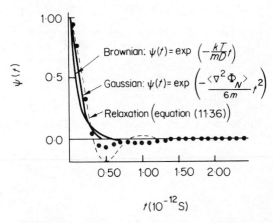

Figure 11.6 Velocity autocorrelations calculated on the basis of the Gaussian model and equation (11.36). This latter equation yields a negative backscattered region in qualitative agreement with the liquid-argon machine calculation. The Gaussian model shows no particle correlation and therefore develops no negative region

easily obtain the criterion for the development of a backscattered negative region for $\psi(t)$—it is simply the condition $\beta^2 < 4\gamma$, or alternatively that the structural relaxation proceeds more slowly than the momentum relaxation. If we set $\gamma \gg \beta$ we obtain a strongly oscillatory velocity autocorrelation more appropriate to a solid: thus large values of γ tend to emphasize the vibratory aspect of the motion in the spectral distribution which is given as the Fourier transformation of equation (11.36). Such a transform is shown qualitatively in Figure 11.7 for the cases $\beta^2 > 4\gamma$ and $\beta^2 < 4\gamma$.

262

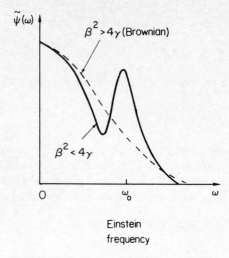

$\tilde{\psi}(\omega)$

$\beta^2 > 4\gamma$ (Brownian)

$\beta^2 < 4\gamma$

O ω_0 ω

Einstein
frequency

Figure 11.7 Spectral densities on the basis of the velocity autocorrelation (11.36)

References

C. A. Croxton, *LSP*.

G. H. A. Cole, *An Introduction to the Statistical Theory of Classical Simple Dense Fluids*, Pergamon (1967), Chapters 8, 9.

CHAPTER 12

Irreversibility

Introduction

One of the major outstanding problems of liquid state physics is the reconciliation of thermodynamic irreversibility with the known dynamic reversibility of the molecular evolution. Why is it that an initially non-equilibrium system will always evolve towards equilibrium where, subject to external constraints, it will remain? Moreover, how is this to be explained in terms of the mechanically reversible equations of motion?

For dilute impulsively interacting systems, Boltzmann was able to demonstrate that the invocation of the hypothesis of 'molecular chaos', according to which the previous dynamical history of a particle is 'forgotten' upon collision, enabled the system to progress irreversibly towards a time independent state, the phase of the representative point at all times satisfying the mechanically reversible Hamiltonian equations of motion. This inability of the representative point to retrace its path guaranteed irreversibility in time. For dense systems in continuous interaction the strong dynamical and configurational coupling over extended regions of space–time clearly rendered the molecular-chaos hypothesis inappropriate. Kirkwood, however, demonstrated that time-averaged, or 'time-smoothed' segments of the phase evolution could be compared and shown to be statistically independent, provided the period over which the smoothing occurred was greater than the temporal coherence of the correlation. This, in fact, is the analogue of the molecular chaos hypothesis of Boltzmann, and as such offers a qualitative explanation of the irreversible approach of a strongly coupled system to a stationary state, subject as usual to any externally applied constraints.

Where, then, has the information gone? Surely if it has not disappeared there is no reason why it should not be dynamically recoverable. The principal thermodynamic functions of state such as pressure, density and so on, upon which we base our observations of irreversibility, are determined mainly by the lower-order distribution functions as we have seen in the earlier chapters on equilibrium properties of dense fluids. The information concerning the initial non-equilibrium condition is rapidly communicated through the inter-particle interaction to higher- and higher-order correlations, and these higher-order correlations condition the thermodynamically important lower-order functions only through a phase-averaging process, so that for example

$$f_{(2)}(\mathbf{p}_1, \mathbf{p}_2, \mathbf{q}_1, \mathbf{q}_2) = \frac{1}{(N-2)!} \iint f_{(3)}(\mathbf{p}_1, \mathbf{p}_2, \mathbf{p}_3, \mathbf{q}_1, \mathbf{q}_2, \mathbf{q}_3) \, d\mathbf{p}_3 \, d\mathbf{q}_3 \quad (12.1)$$

and so on through the familiar hierachical relation up to $f_{(N)}$. The information is not lost therefore—merely irretrievably redistributed such that it cannot be dynamically reversed to yield the specific initial non-equilibrium configuration.

This is not to say that in the course of the spontaneous fluctuations occurring in the phase distribution the initial phase cannot *in principle* be recovered: it can, if we are willing and able to wait long enough. The time we have to wait will, of course, depend upon how closely it must return to its initial phase with a corresponding decrease in the entropy. The more closely we want the initial state to recur the longer we shall have to wait. Poincaré argued that the phase trajectories are ultimately reversible, provided that they can find their way back through the chain of averages, and the initial nonequilibrium configuration will recur, if we wait long enough.

Smoluchowski calculated the Poincaré recurrence time for a fluctuation of 1 per cent of the mean density of oxygen gas contained in a sphere of radius R. The recurrence times at a mean particle density of 3×10^{-19} cm^3 at 300 K are shown for various R in Table 12.1. The recurrence of initial phases is seen

Table 12.1. Recurrence of initial phases

R (cm)	Recurrence time (s)
10^{-5}	$\sim 10^{-11}$
$2 \cdot 5 \times 10^{-5}$	~ 1
3×10^{-5}	$\sim 10^6$
5×10^{-5}	$\sim 10^{17}$
1 cm	$\sim 10^{152}$
Present age of universe	$\sim 10^{17}$ s

not to be impossible, but certainly improbable, and in this sense the second law of thermodynamics—the principle of increase in entropy—tells us not what *must* happen, but what very probably *will* happen. So the apparent conflict between thermodynamic irreversibility and dynamical reversibility for macroscopic systems arises simply because our observations are restricted to infinitesimally short periods in comparison with the Poincaré time.

The H-theorem and the Approach to Equilibrium

Underlying the approach to equilibrium of course, is the process of maximization of the entropy, and this has to be regarded as the driving force in the evolution. We imagine entropy to be produced in a system in which an irreversible process is taking place, reaching a steady state condition consistent with any external constraints. What we have to show is that equilibrium in phase is characterized by a maximum in the entropy, and that once attained the system, but for small statistical fluctuations, *remains* in equilibrium (Figure 12.1), at least for the duration of the Poincaré period.

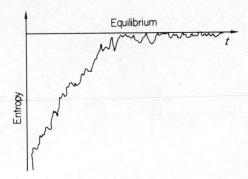

Figure 12.1 Evolution of the entropy with time in an equilibrating system

Such a situation has been demonstrated for dilute gases in which we invoke the assumption of molecular chaos. Boltzmann defined the quantity

$$H(t) = \int f_{(N)} \ln f_{(N)} \, d\mathbf{p}^N \tag{12.2}$$

in terms of the distribution function $f_{(N)}(\mathbf{p}^N, \mathbf{q}^N, t)$. It is straightforward to show that using for $f_{(N)}$ the equilibrium Maxwell–Boltzmann velocity distribution, then $H = -S/k$, where S is the entropy per unit volume of a monatomic ideal gas.

The time differential of (12.2) is

$$\frac{dH(t)}{dt} = \int \frac{\partial f_{(N)}(\mathbf{p}^N, t)}{\partial t} [1 + \ln f_{(N)}(\mathbf{p}^N, t)] \, d\mathbf{p} \tag{12.3}$$

where, to simplify the discussion, we have assumed that the distribution is independent of position and that no external forces act. Moreover, if we assume that $f_{(N)}(\mathbf{p}^N, t) = N f_{(1)}(\mathbf{p}, t)$ which asserts the dynamical independence of the particles it may be shown that (see *LSP*)

$$\frac{dH}{dt} = -\tfrac{1}{4} \iiint \sigma |\mathbf{p}_2 - \mathbf{p}_1| \{f'(1)f'(2) - f(1)f(2)\}$$

$$\times [\ln f(1) \ln f(2) - \ln f'(1) \ln f'(2)] \, d\mathbf{p}_1 \, d\mathbf{p}_2 \, d\mathbf{p}_1' \, d\mathbf{p}_2' \tag{12.4}$$

where σ is the symmetric scattering cross-section between the states $f(1)f(2) \leftrightarrow f'(1)f'(2)$.

It is easily demonstrated that for any x, y

$$(\ln y - \ln x)(y - x) \geqslant 0 \tag{12.5}$$

(the equality holding when $x = y$), and it follows from (12.4) that generally

$$\frac{dH}{dt} \leqslant 0 \tag{12.6}$$

266

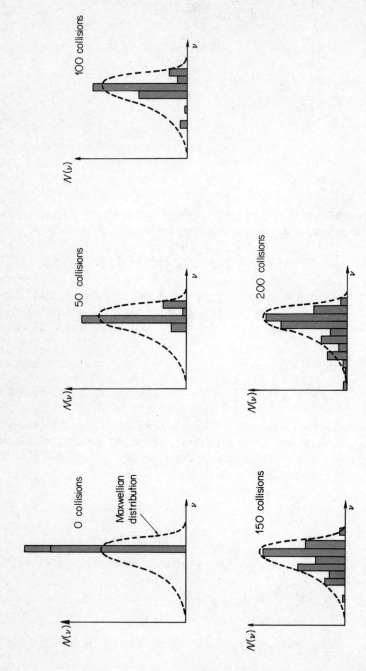

Figure 12.2 Relaxation of an initial non-equilibrium velocity distribution to one of Maxwellian form as a function of number of collisions

which is a statement of the Boltzmann H-theorem. The entropy of a system is of course only defined for equilibrium assemblies: nevertheless for systems not too far from equilibrium (12.6) asserts the increase of entropy as the system approaches equilibrium, maximizing when $dH/dt = 0$, i.e. when

$$f'(1)f'(2) = f(1)f(2) \qquad (12.7)$$

Physically this represents the condition when there is no net population or depopulation of the elementary phase volumes. The equilibrium condition (12.7) is equivalent to

$$\ln f'(1) + \ln f'(2) = \ln f(1) + \ln f(2) \qquad (12.8)$$

This equilibrium condition can only be satisfied in terms of conservation of the three momentum components and the kinetic energy by an expression of the form

$$\ln f(1) = A + B_x mv_x + B_y mv_y + B_z mv_z + C(\tfrac{1}{2}mv^2) \qquad (12.9)$$

where A, B_x, B_y, B_z and C are constants. It then follows that f must be the Maxwellian velocity distribution.

Molecular Dynamics Simulation

Whilst there is no possibility of following the microscopic evolution of a non-equilibrium system experimentally, we may follow the decay of the Boltzmann H-function by means of a molecular dynamics simulation.

If we start a system of 100 hard spheres off from a cubic configuration, in random directions but all with the same velocity, the number and energy of the particles is of course conserved, but the *distribution* evolves as shown in Figure 12.2, finally attaining a Maxwellian form. The H-function may be calculated at the same time and is seen from Figure 12.3 to agree quite closely

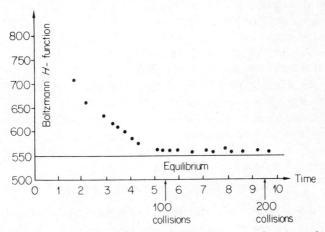

Figure 12.3 Evolution of the H-function as a function of number of collisions

268

with the theoretical equilibrium based on equation (12.4). The slightly higher equilibrium H-value for the 100-particle system may be attributed to the non-development of a high-velocity tail to the distribution factor—this arising from the small molecular sample. Equilibrium is seen to be rapidly attained in about two collisions per particle. If we simulated a more realistic system, of square-well molecules for example, the equilibration is much slower. This is understood in terms of configurational impedance of the total energy distribution: the system has to exchange potential and kinetic energy in coming to equilibrium, and this is governed by configurational adjustment—a problem which does not arise in the hard-sphere system.

Irreversibility at Liquid Densities

We have already observed that the essentially stochastic basis of Brownian motion provides an inadequate description of the dynamical evolution in a dense liquid. Nevertheless, it is clear that Brownian motion, in as far as it consists of an uncorrelated chain of evolutionary events, should show a similarly irreversible progression towards equilibrium as that outlined for a dilute gas in the previous section, and Kirkwood's time-smoothing procedure enables us to account qualitatively for irreversible phenomena in liquids. But what we require is a *general H*-theorem which would show the same irreversible evolution but which nevertheless incorporates the space–time correlation in phase which undoubtedly exists at liquid densities. Since the irreversibility is attributed to a finite memory of the system, some theoretical effort has gone into specifying a 'memory function', but whilst these theories yield velocity autocorrelations and spectral densities in qualitative agreement with experiment, they are not sufficiently *a priori* to allow much insight into the more difficult problem of phase irreversibility in dense, strongly coupled systems of particles.

The concept of a memory function represents, of course, no more than a picturesque description of the irretrievable loss of dynamical information through space–time coupling to higher and higher distribution functions, with the progressive dispersion of information until ultimately the whole system is spanned by a multiplicity of incoherent correlations.

A general theory of interacting particles has been recently formulated, but as yet is applicable only to dilute systems—results at liquid densities are still of a formal nature. Nevertheless, the development of irreversible features may be identified, and whilst this represents no more than an intermediate stage in the general theoretical development of irreversibility, any future theory will undoubtedly have to incorporate the formal results we have already.

General Theory of Interacting Particles

The Liouville equation describing the N-body phase evolution

$$\frac{\partial f_{(N)}}{\partial t} + \sum_{j=1}^{N} \left\{ \frac{\mathbf{p}_j}{m} \frac{\partial f_{(N)}}{\partial \mathbf{q}_j} + \mathbf{F}_j \frac{\partial f_{(N)}}{\partial \mathbf{p}_j} \right\} = 0 \qquad (12.10)$$

may be written in the form

$$\frac{i\partial f_{(N)}}{\partial t} = \mathscr{L} f_{(N)} \tag{12.11}$$

where

$$\mathscr{L} = -i \sum_{j=1}^{N} \left\{ \frac{\mathbf{p}_j}{m} \frac{\partial}{\partial \mathbf{q}_j} + \mathbf{F}_j \frac{\partial}{\partial \mathbf{p}_j} \right\}; \qquad i = \sqrt{-1} \tag{12.12}$$

We observe the formal similiarity between the Liouville equation in operator form (12.11) and the time-dependent Schrödinger equation. This circumstance has been extensively exploited by Prigogine and his school in the discussion of irreversibility using perturbation techniques familiar from quantum field theory. We notice that the Liouville operator $\mathscr{L}$ may generally be written as a sum of a 'non-interacting' particle operator $\mathscr{L}_0$, and $\delta\mathscr{L}$, regarded as a perturbation, which incorporates the effects of atomic interaction and scatters the particle from an initial to a final state:

$$\mathscr{L} = \mathscr{L}_0 + \lambda \delta\mathscr{L} \tag{12.13}$$

λ is a time-independent constant representing the strength of the interaction. In the free (non-interacting) particle case ($\lambda = 0$) solution of (12.11) with $\mathscr{L} = \mathscr{L}_0$ yields a complete set of oscillatory eigenfunctions as solution, characterized by a wavenumber $\mathbf{k}$. The amplitudes of the various eigenfunctions $\rho_{\mathbf{k}}$ are, of course, time-independent for if there is no interaction there can be no evolution, and the amplitudes remain constant in time. For conditions not too far removed from equilibrium we may, as usual, express the general time-dependent evolution as a linear combination of the unperturbed eigenfunctions of (12.11). Progression towards equilibrium may then be discussed in terms of the evolution of the amplitudes $\rho_{\mathbf{k}}$, and whilst at short times the dynamical reversibility is apparent, in the limit of long times the development of the amplitudes $\rho_{\mathbf{k}}$ is seen to consist of a 'cascade' process and an irreversible flow of correlation amongst the coefficients develops.

The Liouville equation (12.11) has the formal solution

$$f_{(N)}(\mathbf{p}^N, \mathbf{q}^N, t) = \exp(-i\mathscr{L}t) f_{(N)}(\mathbf{p}^N, \mathbf{q}^N, 0) \tag{12.14}$$

where $\exp(-i\mathscr{L}t)$, designated the *propogator*, links the phase distributions at times 0 and t.

We shall not go through a detailed mathematical analysis here—for that the reader is directed to *LSP* or Prigogine's monograph—what we shall do however, is proceed directly to the *diagrammatic representation* of the evolution.

The transition from state $\mathbf{k}$ to $\mathbf{k}'$ occurs, of course, in the field of the scattering potential and it is not difficult to show that corresponding to a Fourier component ($\exp \pm i\mathbf{m}\mathbf{q}$) in the scattering potential, there exists the possibility of the transition $\mathbf{k} \to \mathbf{k} \pm \mathbf{m}$. *Non-diagonal* transitions of the kind $\mathbf{k} \to \mathbf{k}'$ represent inelastic interactions between particles whilst *diagonal* transitions $\mathbf{k} \to \mathbf{k}$

correspond to elastic exchanges during which, however, transient excitations may nevertheless occur.

For weak interactions we may take it that m adopts the values ± 1 only. This naturally raises the question of the density of states, and if the evolution is to proceed classically we must assume a *continuum* of states. We are immediately restricted to the condition $V \to \infty$, $N \to \infty$, $N/V \to$ constant: this again raises questions concerning the validity of the small molecular dynamic or Monte Carlo ensembles.

As we pointed out, the conditions $\mathbf{m} = \pm 1$ restricts the transitions which may occur: for example it is not possible to effect the transition $\rho_{\{0\}}(0)$ to $\rho_{\{0\}}(t)$ in an odd number of steps, i.e. (reading from right to left)

$$\langle \{0\} | \delta \mathcal{L} | \{k''\} \rangle \langle \{k''\} | \delta \mathcal{L} | \{k'\} \rangle \langle \{k'\} | \delta \mathcal{L} | \{0\} \rangle = 0 \qquad (12.15)$$

since the first two single-step transitions cannot be compensated by the third. This means that in the dilute weak-coupling limit *only diagrams having an even number of transition vertices need be considered*:

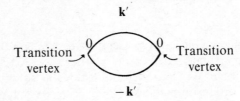

Higher-order elastic or diagonal transitions through more intermediate states may arise

and so on, subject of course to conservation of wave vectors and the weak coupling condition $\mathbf{m} = \pm 1$. The probability of inducing transitions depends upon the strength of the interaction parameter, λ. Thus $\bigcirc$, involving two transitions, is a diagram of order λ^2, $\bigcirc\!\!\bigcirc\!\!\bigcirc$ is of order λ^4, etc., the order being given by the number of vertices. Clearly, in the weak coupling limit the higher-order diagrams make a correspondingly smaller contribution to the evolution.

Now, it may be shown that it is possible to represent the evolution of $\rho_{\{0\}}(t)$ as

$$\rho_{\{0\}}(t) = \rho_{\{0\}}(0) \sum_{n=0}^{\infty} \mathbf{O}_0^n \frac{(\lambda^2 t)^n}{n!} \qquad (12.16)$$

where the operator $\mathbf{O}_0^n$ is given as the sum over the diagonal transition chains

$$\mathbf{O}_0 \equiv \left\{ \lambda^2 \bigcirc + \lambda^4 \bigcirc\!\!\bigcirc\!\!\bigcirc + \lambda^6 \bigcirc\!\!\bigcirc\!\!\bigcirc\!\!\bigcirc\!\!\bigcirc + \cdots \right\}$$

$$(12.17)$$

It is clear that the master equation describing the evolution of the amplitude $\rho_{\{0\}}$ is (cf. equation (12.14))

$$\rho_{\{0\}}(t) = \rho_{\{0\}}(0) \exp\{\lambda^2 t \mathbf{O}_0\} \tag{12.18}$$

or alternatively

$$\partial \rho_{\{0\}}/\partial t = \lambda^2 \mathbf{O}_0 \rho_{\{0\}}(t) \tag{12.19}$$

whereupon it is immediately evident that (12.19) describes a Markovian process in which the Fourier amplitude $\dot{\rho}_{\{0\}}(t)$ is proportional to its current value, and, moreover, is independent of its previous history. The evolution is correspondingly asymmetric in time and irreversible.

We may extend the above analysis to include strongly coupled dilute systems by removing the transition condition $\mathbf{m} = \pm 1$. Then terms containing odd numbers of transition vertices of order λ^3, λ^5, etc. will appear in (12.17). An irreversible evolution of the form (12.19) will develop again of course.

The more general non-diagonal (inelastic) transitions are not so easy to analyse, although some formal results are available, but there are virtually no results available as yet at liquid densities. Nevertheless, this generalized approach to irreversible phenomena in dense strongly coupled systems is actively under investigation and may ultimately yield results which are other than purely formal.

Irreversibility as a Symmetry-breaking Process

A basic feature of the Liouville equation written in operator form

$$\frac{i\partial f_{(N)}}{\partial t} = \mathscr{L} f_{(N)} \tag{12.20}$$

is its '$\mathscr{L}t$ invariance'. In other words under the combined effect of the two operations

$$\begin{cases} \mathscr{L} \rightarrow -\mathscr{L} \\ t \rightarrow -t \end{cases}$$

equation (12.20) remains invariant, and this expresses the time-reversal symmetry. We now make a general enquiry as to how the symmetry may be broken so that there is a privileged direction in time corresponding to an irreversible evolution, and increase of entropy with time. What then are the conditions which have to be imposed on the dynamics to obtain this kind of behaviour?

We may answer this question by noting that equation (12.20) has the formal solution

$$f_{(N)}(t) = f_{(N)}(0) \exp(-i\mathscr{L}t)$$

$$= \frac{1}{2\pi i} \int_c \exp(-izt) \frac{1}{\mathscr{L} - z} f_{(N)}(0)\, dz \tag{12.21}$$

Equation (12.21), known as Bromwich's integral taken along a contour c in the complex plane, represents the formal solution in *resolvent operator* form. The general expression of the resolvent $(\mathscr{L} - z)^{-1}$ is made in terms of creation and destruction operators, but for diagonal transitions the master equation takes a particularly simple form

$$\frac{i\partial \rho_{\{0\}}(t)}{\partial t} = \Psi_0 \rho_{\{0\}}(t) \tag{12.22}$$

where the elastic collision operator adopts the form

$$\Psi_0 = -\langle\{0\}|\mathscr{L}|\{k\}\rangle \frac{1}{\langle\{k\}|\mathscr{L}|\{k\}\rangle - z}\langle\{k\}|\mathscr{L}|\{0\}\rangle \tag{12.23}$$

$$\equiv \langle 0\mathscr{L}k\rangle \frac{1}{\langle k\mathscr{L}k\rangle - z}\langle k\mathscr{L}0\rangle \tag{12.24}$$

In the limit of large systems the sum over intermediate states involves an integral, and a formal representation of (12.24) is given in terms of a principal part and a residue where the resolvent becomes singular:

$$-i\Psi_0 = -\pi\langle 0\mathscr{L}k\rangle\delta(k\mathscr{L}k)\langle k\mathscr{L}0\rangle + i\langle 0\mathscr{L}k\rangle\frac{1}{\langle k\mathscr{L}k\rangle}\langle k\mathscr{L}0\rangle$$

$$= \Psi_0^e + \Psi_0^o \tag{12.25}$$

and we observe that Ψ_0 has an odd Ψ_0^o and even Ψ_0^e part with respect to the inversion $\mathscr{L} \to -\mathscr{L}$ in (12.25). If the even part Ψ_0^e exists then the operations $\mathscr{L} \to -\mathscr{L}, t \to -t$ change the master equation

$$\frac{i\partial \rho_{\{0\}}(t)}{\partial t} = [\Psi_0^e + \Psi_0^o]\rho_{\{0\}}(t) \tag{12.26}$$

into

$$\frac{i\partial \rho_{\{0\}}(t)}{\partial t} = [-\Psi_0^e + \Psi_0^o]\rho_{\{0\}}(t) \tag{12.27}$$

In other words $\mathscr{L}t$ invariance is lost and the evolution is no longer symmetrical in time.

The even part Ψ_0^e is a negative operator so that

$$-i\Psi_0^e \leqslant 0 \quad \text{(dissipativity condition)} \tag{12.28}$$

and therefore expresses a dissipation or damping in the evolution (12.22). Moreover, once satisfied it may be shown that (12.28) leads to an increase in entropy with time. We see that the appearance of an even part in the elastic collision operator Ψ_0 leads to a radical change in the dynamic evolution, and in particular leads to a breakdown in $\mathscr{L}t$ symmetry, establishing a preferred direction in time.

Reference

G. H. A. Cole, *An Introduction to the Statistical Theory of Classical Simple Dense Fluids*, Pergamon (1967), Chapter 7.

INDEX

Page references in *italics* are to tables or figures in which experimental data or the results of theoretical calculations are presented. Chemical elements are listed under their conventional symbols only if they are the subject of special comment in the text.

Abe approximation, and diagrammatic representation 71–73
  relation to BGY equation 71
  relation to HNC equation 73
Ag, compressibility *91*
  Hall coefficient *192*
  surface tension *185*
alkali metals, compressibility *91*
  critical properties *160*, 167
  molecular dynamics 241–243
  pair potentials 241–243
  pseudopotential 197
  resistivity *196*, *199*
  structure factor 84
  surface properties *172*, 183–185
  velocity autocorrelation 243
alloys, resistivity 202–204
  structure 243
ankylosis 5
Ar, anisotropy at liquid surface, kinetic, 236–241
  structural 179–183, 237, *238*
  compressibility *91*
  configurational properties *44*
  critical properties *158*, *160*, 161
  diffusion coefficient 231, 239, *256*
  equation of state,
    BGY 120
    HNC 95, 120–122, 141–145
    Padé approximant 115
    PY 121
    virial 108–111
  friction constant *256*
  machine simulation 136, 229–234
  pair potential 6–9, 31, *33*, 124
  phase transition 136, 141–146
  specific heat 57, 161
  spectral density 57, 233
  surface energy 190
  surface structure 179, 182, *183*, 189
  surface tension 190
  thermodynamic properties 95

Ar—*continued*
  velocity autocorrelation 233, 239, 261
  virial coefficients 31–34
autocorrelation force 254–257
  relation to friction constant 253–254
autocorrelation velocity 232–234, 239, 260–262
  relation to diffusion coefficient 234

B_n, *see* virial coefficients
backscattering 233
bend, of director field 210–211, 213
BGY equation, *see* Born–Green–Yvon equation
Bi, compressibility 91
  Hall coefficient *192*
  resistivity *199*
bifurcation equation 141
binary encounters 248–251
biological membrane, phase transition in 220–222
blip function 101
blurring of Fermi surface 100, 203
Boltzmann transport equation 248
Born–Green–Yvon equation 125
  comparison with other theories 111–124
  inversion of 122–124
  Padé extension of 67–70
  phase transition 137–141
  physical derivation of 108
  stability of phase 138
Bose–Einstein statistics 35 *et seq.*
boson fluid 32, 35, 39, 128, 235
boundary conditions, periodic 224
  weak and strong (Kirkwood and Eisenschitz) 259
Boyle temperature 30, 32
bridge diagrams, *see* elementary diagrams
Brillouin zone 224
Brownian motion 230, 252, 260–262
bulk modulus, *see* compressibility
bundle diagrams *78*